Greater Sage-Grouse National Research Strategy

By Steven E. Hanser and Daniel J. Manier

Scientific Investigations Report 2013–5167

U.S. Department of the Interior
U.S. Geological Survey

U.S. Department of the Interior
SALLY JEWELL, Secretary

U.S. Geological Survey
Suzette M. Kimball, Acting Director

U.S. Geological Survey, Reston, Virginia: 2013

For more information on the USGS—the Federal source for science about the Earth, its natural and living resources, natural hazards, and the environment, visit http://www.usgs.gov or call 1–888–ASK–USGS.

For an overview of USGS information products, including maps, imagery, and publications, visit http://www.usgs.gov/pubprod

To order this and other USGS information products, visit http://store.usgs.gov

Suggested citation:
Hanser, S.E., and Manier, D.J., 2013, Greater Sage-Grouse National Research Strategy: U.S. Geological Survey Scientific Investigations Report 2013-5167, 46 p. plus appendix, http://pubs.usgs.gov/sir/2013/5167/.

Contents

Contents—Continued

Figures

Tables

Greater Sage-Grouse National Research Strategy

By Steven E. Hanser and Daniel J. Manier

Executive Summary

The condition of the sagebrush ecosystem has been declining in the Western United States, and greater sage-grouse (*Centrocercus urophasianus*), a sagebrush-obligate species, has experienced concurrent decreases in distribution and population numbers. This has prompted substantial research and management over the past two decades to improve the understanding of sage-grouse and its habitats and to address the observed decreases in distribution and population numbers. The amount of research and management has increased as the year 2015 approaches, which is when the U.S. Fish and Wildlife Service (FWS) is expected to make a final decision about whether or not to protect the species under the Endangered Species Act.

In 2012, the Sage-Grouse Executive Oversight Committee (EOC) of the Western Association of Fish and Wildlife Agencies (WAFWA) requested that the U.S. Geological Survey (USGS) lead the development of a Greater Sage-Grouse National Research Strategy (hereafter Research Strategy). This request was motivated by a practical need to systematically connect existing research and conservation plans with persisting or emerging information needs. Managers and researchers also wanted to reduce redundancy and help focus limited funds on the highest priority research and management issues.

The USGS undertook the development of this Research Strategy, which addresses information and science relating to the greater sage-grouse and its habitat across portions of 11 Western States. This Research Strategy provides an outline of important research topics to ensure that science information gaps are identified and documented in a comprehensive manner. Further, by identifying priority topics and critical information needed for planning, research, and resource management, it provides a structure to help coordinate members of an expansive research and management community in their efforts to conduct priority research.

This Research Strategy was developed by the USGS using a four-step process:

1. Research needs, questions, ideas, or uncertainties about sage-grouse populations, sagebrush habitats, and change agents were identified by conducting a thorough review of National and State conservation assessments, plans, and strategies.

2. Research questions were categorized into themes and topics.

3. Topics were prioritized using a focus group made up of representatives from Federal and State agencies.

4. The written report was drafted by USGS staff, followed by colleague and technical review and revision. The review and approval of the final publication was consistent with USGS Fundamental Science Practices (Fundamental Science Practices Advisory Committee, 2011).

Despite being one of the most well-studied upland game birds in North America, key gaps in the knowledge of sage-grouse biology remain, and many of these information needs directly affect management planning and implementation. Filling these gaps can inform future adaptive management in a complex and changing environment. The development of population models that incorporate information about the complexities of the biological processes and dynamic habitats in which sage-grouse occur is a first step. A starting point for this modeling is aggregation of the wealth of existing demographic and population data, followed by analysis of these data using modern statistical tools. In addition, new understanding of links among population connectivity, habitat conditions, and arrangements or patterns of habitat is important for understanding meta-population processes. Sage-grouse genetic analyses would provide the necessary information to describe relatedness among breeding locations, delineate population structure, and describe movements among populations throughout the sage-grouse range. Implementation of a unified approach for monitoring sage-grouse across its range, including the multiple periods of its life cycle and variations in habitat conditions, would be the foundation for future analysis and increase the power of multi-scale assessments of sage-grouse population characteristics and assessments of the response of populations to change.

Identification of effective management practices to maintain or improve sage-grouse habitat depends on understanding the components of sage-grouse habitat and sagebrush ecosystems that facilitate robust sage-grouse populations. Knowledge of direct and indirect links between habitat condition and configuration and sage-grouse demographic processes at multiple scales will help inform site-level habitat maintenance, rehabilitation, and restoration activities, including placement of those activities within the appropriate landscape context. This information can be gained through integrated analyses of restoration practices, ecosystem succession, recovery rates, ecosystem function, and environmental covariates. Understanding the components of habitat suitability that affect the ability of individual sage-grouse to move within and between seasonal habitats, as well as among populations will help define management options to meet sage-grouse life-history needs and maintain population connectivity.

Human actions and natural processes affect sage-grouse and their habitats through a variety of mechanisms. An important step toward understanding these mechanisms is the determination of the effects of habitat loss and alteration due to anthropogenic surface disturbance and related activities on sage-grouse behavior and population characteristics. By emphasizing the relations between habitat conditions and population responses, these studies could evaluate the effectiveness of current management guidelines and practices and identify new alternatives.

Conifer encroachment and spread of invasive species are both natural processes affected by human influences with the potential for large effects on sagebrush and sage-grouse. The effectiveness of habitat treatments to restore functioning sage-grouse habitat in areas where conifer encroachment occurs and the long-term cost-benefit of those actions needs further study. Development of new practices and improvements to existing ones could help reduce or eliminate the spread of invasive plant species, as well as help rehabilitate or restore invaded sagebrush habitats.

Fire is an important natural influence on sage-grouse and sagebrush ecosystems, and concerns about loss of sagebrush habitats have led to a multitude of approaches to control the spatial extent of areas burned and effects of fire when it occurs. Understanding the effects of these fire reduction activities on sagebrush habitat quality and local sage-grouse populations could improve application of these types of efforts in the context of sage-grouse conservation. Further, assessment of fire history and fire-recovery rates in ways that elucidate the role of past disturbance in determining fire patterns and frequencies will inform planning efforts and deployment of resources to achieve multiple goals, including sage-grouse conservation.

The influence of herbivory, including grazing by wild horses (*Equus ferus*), burros (*Equus asinus*), elk (*Cervus canadensis*), deer (*Odocoileus* spp.), pronghorn (*Antilocapra americana*), and domestic livestock, on sage-grouse populations and habitat conditions is an important question that can be addressed through a coordinated multi-agency effort. Scientific information necessary to help refine management practices and diminish negative effects on sage-grouse and sagebrush habitats is needed for local, regional, and range-wide scales.

Addressing the research priorities presented in this strategy in ways that are integrated and complementary requires improved integration of data and expertise among Federal and State agencies and non-government organizations. Repeated requests for more consistent techniques and meta-analyses that aggregate data across regions indicate the desire for cooperative refinement of population estimation methods and development of consistent approaches to link population responses to environmental conditions, including fire, to use by herbivores, to land-use change, to infrastructure development, and to other human activities. Collaborations by the Bureau of Land Management (BLM), U.S. Forest Service (USFS), FWS, and State wildlife agencies have resulted in elaborate, spatially explicit conservation plans for habitat and population management. These approaches and designations (for example, priority habitats) represent a large experiment in landscape management that balances human land-use demand with conservation of sage-grouse and their habitats. The effectiveness of these landscape approaches would benefit from testing and examination to assess outcomes and additional options. Understanding the implications of a complicated and variable landscape, as related to sage-grouse population dynamics, would benefit from a coordinated and interdisciplinary approach beyond typical population or habitat research.

1.0 Introduction

Sagebrush-steppe vegetation historically covered 89 million ha in North America (McArthur and Ott, 1996). Nearly one-half of this habitat was lost or degraded over the last 100 years (Miller and others, 2011), and sagebrush is now considered one of the most imperiled ecosystems in North America (Noss and Peters, 1995). Sage-grouse are sagebrush-obligate species (Patterson, 1952), and sagebrush provides cover and an important component of their diet throughout the year (Connelly and others, 2011a). With declines in sagebrush habitats, populations of greater sage-grouse (*Centrocercus urophasianus*) have decreased in numbers (Garton and others, 2011) and distribution (fig. 1; Schroeder and others, 2004).

Since the late 1990s, the number of peer-reviewed publications that focused on sage-grouse has increased, and this upturn in publication activity is roughly coincident with the first of eight petitions filed between 1999 and 2003 for the listing of greater sage-grouse under the Endangered Species Act. In 2004, the "Conservation assessment of greater sage-grouse and sagebrush habitats" (Connelly and others, 2004) was compiled for the Western Association of Fish and Wildlife Agencies (WAFWA). The goal was to inform the 2005 range-wide listing decision for greater sage-grouse by the U.S. Fish and Wildlife Service (FWS; U.S. Department of the Interior, 2005). In 2004, the conservation assessment was

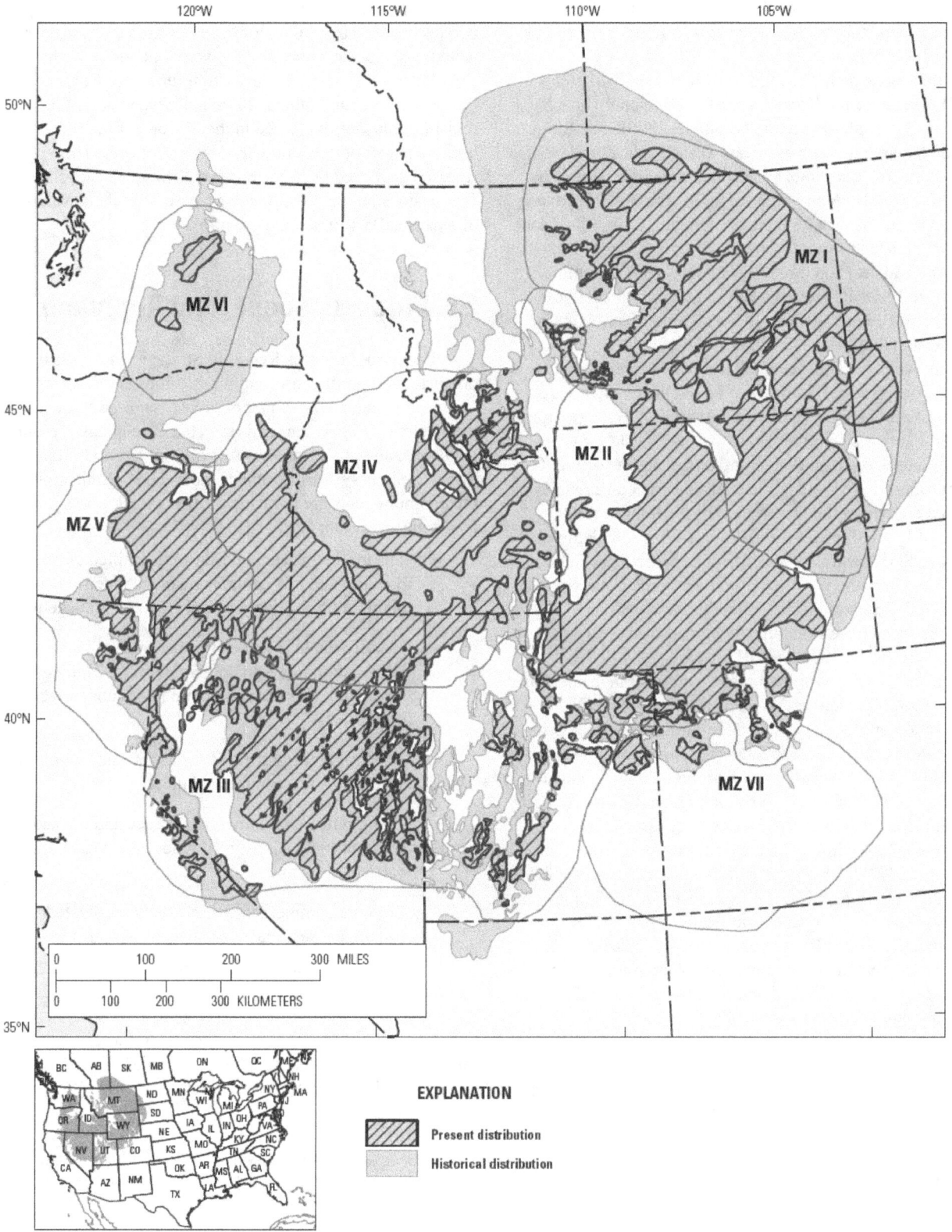

Figure 1. Present and historical distribution of greater sage-grouse in North America (adapted from Schroeder and others, 2004) within Western Association of Fish and Wildlife Agencies designated management zones (Stiver and others, 2006).

the most comprehensive synthesis of information about sage-grouse biology, habitat requirements, and functioning of the sagebrush ecosystem to date.

Other assessments have followed. Markedly in 2006, WAFWA released the Greater Sage-Grouse Comprehensive Conservation Strategy (Stiver and others, 2006), which identified a large number of threats to sage-grouse and created a list of research questions to be addressed. To accomplish many of the goals identified in the 2006 conservation strategy, the 11 Western States (Washington, Oregon, California, Idaho, Nevada, Utah, Montana, Colorado, Wyoming, North Dakota, and South Dakota) and three provinces (British Columbia, Alberta, and Saskatchewan) where sage-grouse occur, joined with several Federal agencies in the signing of a Memorandum of Understanding (MOU) in 2008.

This MOU established the Range-wide Interagency Sage-Grouse Conservation Team (RISCT) and the Greater Sage-Grouse Executive Oversight Committee (EOC). These groups brought together executive (EOC) and technical (RISCT) staff from each organization to implement conservation of greater sage-grouse and their habitat. The Federal agencies involved are within the Department of the Interior [DOI; Bureau of Land Management (BLM), FWS, U.S. Geological Survey (USGS)] and the Department of Agriculture [USDA; U.S. Forest Service (USFS), Natural Resource Conservation Service (NRCS), and Farm Service Agency (FSA)].

States also have been organizing and conducting activities. Between 2003 and 2011, each State wildlife management agency within the sage-grouse's range developed conservation plans and assessments to inform statewide conservation efforts (Wyoming Sage-Grouse Working Group, 2003; Nevada Sage-Grouse Conservation Team, 2004; Stinson and others, 2004; McCarthy and Kobriger, 2005; Montana Sage Grouse Work Group, 2005; Idaho Sage-Grouse Advisory Committee, 2006; Colorado Greater Sage-Grouse Steering Committee, 2008; South Dakota Department of Game, Fish, and Parks, 2008; Utah Division of Wildlife Resources, 2009; Hagen, 2011a). A scientific monograph about the ecology and conservation of greater sage-grouse and its habitats (Knick and Connelly, 2011) also was developed by Federal, State, and non-governmental personnel. The monograph revised information in Connelly and others (2004) and provided new information not addressed in the 2004 assessment. These documents formed a base of knowledge used for the 2010 decision by the FWS that listing sage-grouse range-wide was warranted under the Endangered Species Act but precluded by other priorities (U.S. Department of the Interior, 2010). Subsequent to this 2010 decision, settlement agreements between the FWS and environmental groups established a schedule for a final decision by September 30, 2015. This timeline and associated conservation concerns have increased requests for information about sage-grouse populations and their habitats.

Despite the amount of research conducted over the past two decades, a large number of key research questions remain unaddressed or only partially answered. In addition, the 2006 WAFWA Greater Sage-Grouse Comprehensive Conservation Strategy (Stiver and others, 2006) anticipated revisiting and revising priorities expressed in that document in 2012. It became apparent that a new framework for organizing priority research was needed. With the approaching FWS decision, the EOC requested that USGS take a lead role in the development of a national research strategy.

2.0 Purpose, Scope, and Approach

The purpose of this Research Strategy is to identify, organize, and outline information priorities and research needs to improve the understanding of greater sage-grouse and their habitats[1]. This strategy is targeted at scientists, resource managers, and policy-makers working for Federal, State, Tribal, and non-governmental agencies and organizations. Overarching goals are to provide a framework for sage-grouse research in order to:

- Ensure that research needs are identified and documented in a comprehensive manner to support planning, prioritization, and resource management;

- Identify the highest priority research and science information needs to help guide use of limited financial resources and eliminate redundancy in efforts;

- Expand partnerships and collaborations within the sage-grouse research community;

- Communicate broadly with science and management communities to inform and motivate widespread participation; and

- Promote complementary and (or) coordinated research at regional and range-wide scales.

The Research Strategy addresses sage-grouse life history, habitat management, natural disturbance, and influences of human actions and infrastructure on sagebrush systems and sage-grouse. In many cases, the issues addressed are interrelated, which means inclusive or multifaceted approaches have particular merit. Specific objectives include:

---

[1]Some of the language of this report describes advantageous or suitable research to support understanding and management decisions. This is done with recognition that many factors besides the evaluations described or cited in this Research Strategy may eventually come to bear in planning and conducting research. Explicit directives or judgments are not intended.

- Identify and characterize the range of questions and issues expressed by managers and researchers relating to the management and conservation of sage-grouse and sagebrush ecosystems; and

- Prioritize research topics according to management need.

A phased approach was used to develop the Research Strategy. First, a review was conducted of the National and State conservation assessments, plans, and strategies (table 1) to identify research needs, questions, ideas, or uncertainties about sage-grouse populations, sagebrush habitats, and threats to either or both. Then research needs were categorized into three broad themes—sage-grouse biology, sage-grouse habitat management, and change agents. These themes contain multiple topics, and at least one research question was associated with each topic (appendix A). Sources of

original topics and questions were preserved in a database for reference and to help identify patterns and variations in priorities. Prioritization was accomplished using a focus group of representatives from Federal and State agencies (table 2). The initial expectation was that an explicit ranking of research needs would emerge, but the complexity and breadth of the issues resulted in intermixing of topics, disparities among regions, local-to-regional incompatibilities, and differences in opinion among experts. This resulted in thematic categories being developed using a matrix approach for coordinated or interrelated efforts[2].

---

[2]Detailed narratives are presented in section 4.0, "Research Themes and Topics." The narratives include a discussion of the research topics that represent the issues and questions used to identify and characterize strategic themes and topics. These topics and narratives provide a synthesis of research needs, but are not an exhaustive list.

Table 1. Conservation and management documents reviewed to identify research needs and management issues.

Citation	Year	Title
Wyoming Sage-Grouse Working Group	2003	Wyoming greater sage-grouse conservation plan
Nevada Sage-Grouse Conservation Team	2004	Greater sage-grouse conservation plan for Nevada and eastern California
Stinson and others	2004	Washington state recovery plan for the greater sage-grouse.
Montana Sage Grouse Work Group	2005	Management plan and conservation strategies for sage grouse in Montana – Final
McCarthy and Kobriger	2005	Management plan and conservation strategies for greater sage-grouse in North Dakota
Stiver and others	2006	Greater sage-grouse comprehensive conservation strategy
Idaho Sage-Grouse Advisory Committee	2006	Conservation plan for the greater sage-grouse in Idaho
South Dakota Department of Game, Fish, and Parks	2008	Greater sage-grouse management plan South Dakota 2008–2017
Colorado Greater Sage-Grouse Steering Committee	2008	Colorado greater sage-grouse conservation plan
Utah Division of Wildlife Resources	2009	Utah greater sage-grouse management plan
Stiver and others	2010	Sage-grouse habitat assessment framework
Hagen	2011a	Greater sage-grouse conservation assessment and strategy for Oregon: A plan to maintain and enhance populations and habitat
Knick and Connelly	2011	Greater sage-grouse: Ecology and conservation of a landscape species and its habitats
Sage-grouse National Technical Team	2011	A report on national greater sage-grouse conservation measures
U.S. Department of the Interior	2012	Sage-grouse conservation objectives draft report
Range-wide Interagency Sage-Grouse Conservation Team	2012	Near-term greater sage-grouse conservation action plan
Manier and others	2013	Summary of science, activities, programs and policies that influence the rangewide conservation of greater sage-grouse

Table 2. Participants in the review and prioritization meeting held in Boise, Idaho, March 4–5, 2013.

Name	Agency
Aaron Robinson	North Dakota Game and Fish Department
Cameron Aldridge	Colorado State University and U.S. Geological Survey
Christian Hagen	Oregon State University
Clint McCarthy	U.S. Forest Service
Jack Connelly	Idaho Department of Fish and Game
Mike Schroeder	Washington Department of Fish and Wildlife
San Stiver	Western Association of Fish and Wildlife Agencies
Shawn Espinosa	Nevada Department of Wildlife
Steve Knick	U.S. Geological Survey
Sean Finn	U.S. Fish and Wildlife Service
Tom Christiansen	Wyoming Game and Fish Department
Tom Rinkes	Bureau of Land Management

3.0 Strategic Research Approach

Agencies and individuals interested in sage-grouse and sagebrush ecosystems have specific missions and goals associated with their organizational contexts. This Research Strategy identifies important areas for emphasis or collaborative involvement that respects those contexts as well as provides common ground for actions and understanding. The research priorities provide a framework for expanding scientific understanding of sage-grouse biology and the sagebrush ecosystem to inform the management needed to maintain or restore sage-grouse habitat and conserve populations (see "Strategic Sage-Grouse Research Framework"). The organizational structure of the strategy is simple. Priority research topics are categorized in three themes: sage-grouse biology, habitat management, and change agents. Sage-grouse biology addresses topics that inform adaptive management of populations dependent on a complex, changing environment. Habitat management addresses topics that foster conservation and management of functioning sage-grouse habitat and sagebrush ecosystems. Change agents addresses topics that incorporate awareness and consideration of the factors associated with major changes in the sagebrush ecosystem, including their mechanisms, synergies, linkages, and effects on sage-grouse. Research within each of these topics and themes may be considered of equal importance, and it is recognized that different organizations will become involved in addressing the topics and themes to varying degrees based on their respective missions and resources.

Cutting across the prioritized topics are several important, unifying issues that help to characterize a research strategy that supports regional and range-wide conservation efforts. Methods and analytical approaches are needed that can efficiently and appropriately compile data across regions to assess the effects of spatial and temporal variation in environmental conditions on sage-grouse populations and their habitats. Importantly, this includes standardization of methods for collecting new data so that future analyses can benefit from the consistency and reduced variability in effort across projects and regional programs. Standardization would improve population demographic estimates, including lek counts, nest monitoring, and population productivity estimation, as well as habitat condition monitoring (for example, rangeland condition assessments) particularly in seasonal and disturbed habitats where comparisons among years and across local and regional boundaries are desirable. The establishment of a standardized core set of variables collected on any project addressing sage-grouse biology and habitat conditions would substantially improve the ability of the research community to aggregate data into regional or range-wide datasets. This core-variable approach would recognize the individuality of research projects while providing a mechanism for leveraging the large number of on-going and future site-specific data-collection efforts. The standardization of methods, approaches, and target variables will benefit both management and research by helping to support multi-scale comparative analyses that can lead to an understanding of drivers of sage-grouse population dynamics at the appropriate spatial scales. This outcome would then provide context for local actions and regional planning, and inform adaptive management and assessments of change over time.

Investigation of multiple research topics using consistent methods within a single research project or management application would reduce costs, expedite the process of obtaining results, and help define an integrated understanding of sage-grouse population dynamics and habitat roles and functions. Large landscape assessments have shown that the practice of incorporating multiple research topics into a single project can increase the power of an assessment and provide opportunities unavailable when assessing single questions. For example, incorporating estimates of habitat patterns and conditions, population demographics, and disturbance by human activity in the same project would enhance understanding of relations between sage-grouse and habitat conditions and permit the identification of correlations, synergies, or divergences in these measures that may be important for management. Although population demographics are the ultimate concern of National, State, and local wildlife management and conservation specialists, habitat conditions are the most likely and available target for management. Therefore, better understanding of effects of habitat pattern and condition on sage-grouse habitat selection and population dynamics is essential. Within this context, integration of the influence of human activity on habitat conditions and direct effects on sage-grouse populations can guide decisions, and discerning effects of multiple environmental factors requires an integrated, multivariate approach.

Strategic Sage-Grouse Research Framework

Increase knowledge of sage-grouse biology to inform adaptive management of populations in a complex and changing environment.

- Develop spatially explicit population models that incorporate the complexities of biological processes and dynamic habitats and derive scenarios that reflect local management possibilities (options, opportunities, and obstacles) to support planning decisions.
- Determine links among functional connectivity (intermixing of birds), habitat conditions, and habitat configuration using genetic evidence and sage-grouse movement patterns.
- Develop options for a unified approach for monitoring sage-grouse across its range and within multiple periods of its life cycle.

Understand the components of sage-grouse habitat and sagebrush ecosystems, and identify effective management practices to improve habitat conditions.

- Determine links between multi-scale habitat condition and configuration and sage-grouse population processes.
- Inform site-level management, restoration, mitigation, and rehabilitation activities through integrated study of restoration practices, ecosystem succession, recovery rates, ecosystem function, and environmental covariates.
- Determine the components of habitat suitability that affect the ability of individual sage-grouse to move through the landscape and populations to mix.

Identify factors (change agents) that affect sage-grouse populations and their habitats, and identify management practices that ameliorate negative effects.

- Determine the effects of loss of habitat and the ecological influence of new or altered landscapes due to surface disturbance on sage-grouse behavior and population characteristics.
- Examine the effects of conifer encroachment on sagebrush, and the effectiveness of management treatments to restore functioning sage-grouse habitat in areas where encroachment occurs.
- Develop new practices or improve existing practices to reduce or eliminate the spread of invasive species and to restore sagebrush that is affected by invasive plant species.
- Assess fire history and fire-recovery rates in a way that informs planning efforts and deployment of resources for future fire events, improve the understanding of effective post-fire restoration methods, and link these to sage-grouse population and behavioral data to increase understanding of the response of sage-grouse to fire.
- Determine the influence of herbivory, including grazing by domestic livestock and wild horses (*Equus ferus*), burros (*Equus asinus*), elk (*Cervus canadensis*), deer (*Odocoileus* spp.), and pronghorn (*Antilocapra americana*), on sage-grouse populations and habitat conditions, and develop options that minimize negative outcomes.

While it may be desirable to clearly and simply rank the list of priority topics (for example, those described in Section 4.0, "Research Themes and Topics"), differences in population status, regional land use, regional habitat conditions, and different research histories result in different priorities depending on focus (for example, population versus habitat) and region (for example, Southern Great Basin versus Northern Plains). Integration of methods and concepts across spatial scales is needed in all regions, and specific topics and management issues will dictate differences in the topics that are combined in future, integrated research projects. In addition, sometimes important research is overlooked or is inaccessible to wildlife and habitat-management specialists for various reasons. Improved, ongoing coordination between managers and researchers would mitigate this situation. Managers would ask questions that can be addressed in scientific designs, and scientists could design and conduct research that focuses on managers' questions. This approach would not limit scientists from adding additional questions or otherwise adding perspective to their research beyond specific management questions.

4.0 Research Themes and Topics

This section contains narrative summaries for the multiple research topics identified during the review of Federal and State conservation plans and strategy documents. A hierarchical structure was developed to organize the list of research needs around three broad themes: sage-grouse biology, sage-grouse habitat management, and change agents. Each theme contains multiple topics, and at least one research question is associated with each topic (appendix A). The organizational structure is designed to capture and represent important topic areas, and because of the integrated, correlated, or otherwise interrelated nature of many of the topics, some categories and issues occur repeatedly. This structure may be most valuable to readers when used for topical reference, as opposed to a linear reading progression. Each topic is ranked in priority as either low (L), medium (M), or high (H) based on input of a focus group (tables 3–5). Although it is possible to get a superficial view of research priorities by considering the hierarchy of themes and topics alone, sufficient understanding of the topics to develop relevant research projects requires details about questions and connections described under those headings. Those details are presented here.

4.1 Sage-Grouse Biology

Despite decades of research on sage-grouse life-history attributes, key knowledge gaps persist in the understanding of biological processes (table 3). Population models that incorporate understanding of multiple, complex biological processes with information about the dynamic qualities of sage-grouse habitat would help integrate information about sage-grouse population and habitat dynamics. Consistent and spatially explicit estimation of variables representing population and habitat conditions and dynamics would facilitate comparative analyses and facilitate research and management across the sage-grouse range. Improved techniques and consistent multi-scale applications would support regional population modeling while also providing information about local conditions for immediate planning and implementation of habitat and population management activities. Expanding knowledge about the condition and connectedness of sage-grouse populations across the landscape is a critical component of this effort. This is because variability in the environment and populations across regions and range-wide remains an obstacle to comparative analyses and extrapolation of information to unsampled areas, even though completed studies document reactions of individuals and populations to particular events (for example, a wildfire).

4.1.1 Population Modeling [H]

Relationships between sage-grouse and their habitats vary by season, region, and environmental condition. Managers seek knowledge of how their practices affect sage-grouse locally and regionally, and how they affect sage-grouse in a range-wide context. This level of understanding requires information about sage-grouse population characteristics, such as survival, mortality, and dispersal rates, and requires additional information about how those characteristics change over time and vary by habitat. Sophisticated population models can associate population changes with spatially explicit information about habitat conditions and trends. Habitat models used in these analyses need to address factors that explicitly affect populations, for example land use, as well as chance fluctuations that can result in extinction, such as weather events or disease. These analyses would provide spatial representations of information about conditions that affect the viability of populations over short and long periods of time; for example, the combination of current and potential dynamics in land-use and climate patterns may interact to determine the habitat patterns and productivity in the future. Additionally, habitat conditions will affect individual health and population demographics, which means that all these factors need simultaneous consideration in population models [for example, population viability analyses (PVA)]. Informative modeling approaches will highlight integration of spatially and temporally explicit representations of population and habitat dynamics to define connections and support scenario modeling.

Assessment of population viability is a clear priority, along with several other topics that directly inform these assessments. In addition, parameter estimates that are used in population models have implications for many other topics related to populations and habitat conditions (fig. 2). The importance of integrating demography and population dynamics with spatial and stochastic processes to provide spatially explicit population models was stressed throughout the prioritization process. These types of analyses linking vital rates, metapopulation structure, and dynamic habitat models have been conducted for ovenbirds (*Seiurus aurocapillus*; Larson and others, 2004), blue alcon butterflies (*Maculinea alcon*; Radchuk and others, 2012) and ruffed grouse (*Bonasa umbellus*; Blomberg and others, 2012b), but not for sage-grouse. These analyses require inclusion of spatial variation (for example, rescue effects), temporal variation in underlying processes (both within and among populations), individual- and population-level heterogeneity including differences in survival and reproductive parameters (for example, influence of super females), and deterministic and stochastic changes in habitat and other environmental factors. Importantly,

Table 3.   Research topics in the sage-grouse biology theme and priority designations (low [L], medium [M], or high [H]) based on input from a focus group of representatives from Federal and State agencies.

Topic	Priority designation	Topic No.
Population Modeling	H	4.1.1
Demographics	H	4.1.1.1
Mortality	H	4.1.1.2
Habitat Conditions and Change Agents	H	4.1.1.3
Implications of Priority Areas for Conservation	H	4.1.1.3.1
Movement Patterns and Connectivity	H	4.1.1.3.2
Population Dynamics	M	4.1.1.4
Reproduction	M	4.1.1.4.1
Juveniles	H	4.1.1.4.2
Productivity and Recruitment	M	4.1.1.4.3
Implications of Population Cycles	L	4.1.1.4.4
Genetics Applications and Effective Population Size	H	4.1.1.4.5
Connectivity	H	4.1.2
Movement Patterns	M	4.1.2.1
Habitat	H	4.1.2.2
Barriers and Inhospitable Conditions Between Habitats	H	4.1.2.3
Genetic Evidence	H	4.1.2.4
Population Monitoring	H	4.1.3
Lek Counts	H	4.1.3.1
Demography	M	4.1.3.2
Sex Ratios	H	4.1.3.3
Genetics	H	4.1.3.4
Brood and Juvenile Surveys	M	4.1.3.5
Isolated Populations	M	4.1.3.6
Pellet Counts	L	4.1.3.7
Multiple Scale Relationship and Inference	H	4.1.3.8
Mortality Agents and Factors	H	4.1.4
Sources, Rates, and Influences	M	4.1.4.1
Predation	M	4.1.4.2
Interactions with Habitat Condition	M	4.1.4.2.1
Interactions with Infrastructure	H	4.1.4.2.2
Control Effects	M	4.1.4.2.3
Harvest	M	4.1.4.3
Population Dynamics	M	4.1.5
Demography	H	4.1.5.1
Reproduction	M	4.1.5.2
Productivity and Recruitment	H	4.1.5.3
Isolated Populations	M	4.1.5.4
Behavior	M	4.1.6
Dispersal	H	4.1.6.1
Seasonal Movement Patterns	M	4.1.6.2
Seasonal Habitat Selection	M	4.1.6.3
Food	M	4.1.7
Adaptation	L	4.1.8
Translocation	L	4.1.9

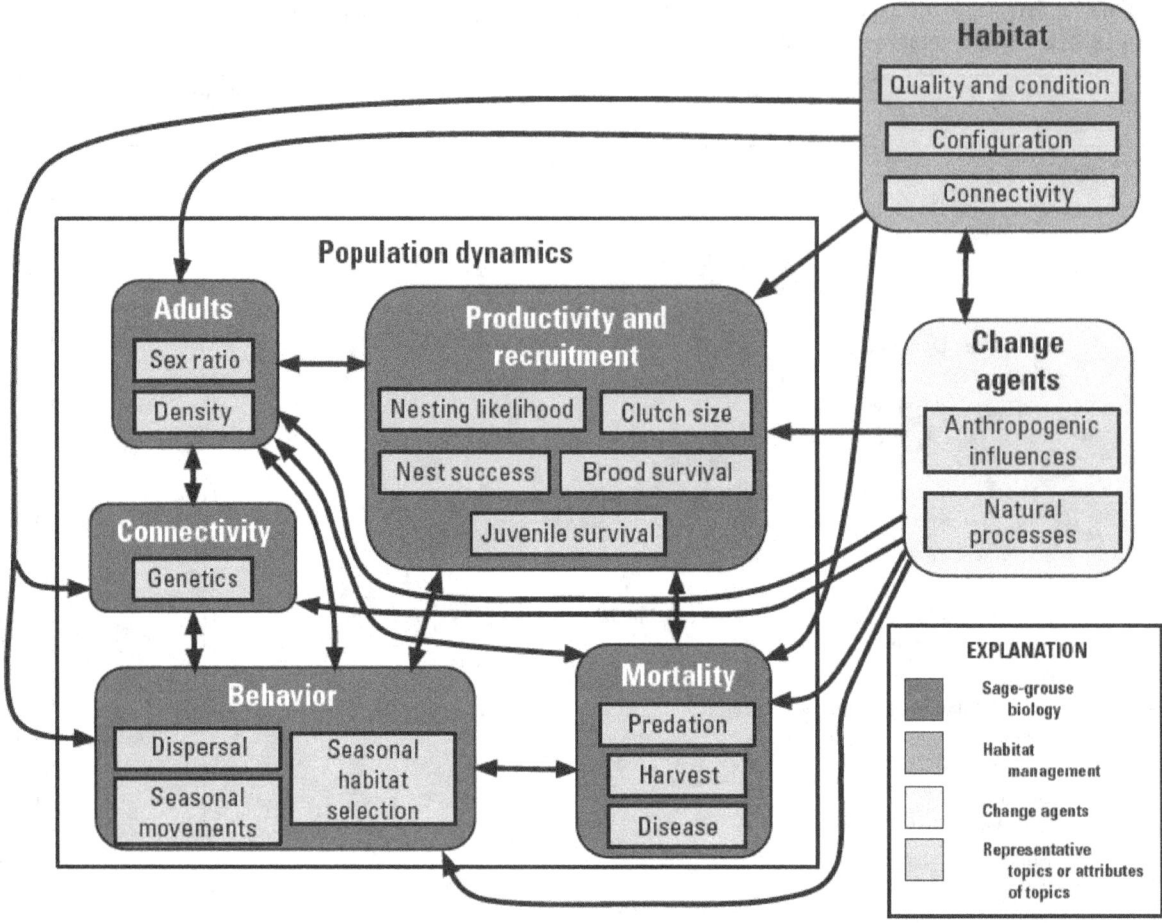

Figure 2. Conceptual model of research topics within the sage-grouse biology theme. Boxes represent research topics or representative attributes within a theme and arrows provide interactions between themes and topics.

population and habitat parameters that are used in population models also are valuable for numerous management and planning activities, such as habitat management, monitoring population dynamics, assessing reproductive success, and specifying the type and timing of mortality.

4.1.1.1 Demographics [H]

Sage-grouse populations have been counted and monitored for decades, yet variations in count methods, dynamic populations, including short- and long-term trends, and variability among populations and environmental covariates, have made regional assessments and population monitoring challenging (for example, Connelly and others, 2003, 2004; Dahlgren and others, 2010a, 2010b; Bruce and others, 2011; Fedy and Aldridge, 2011). Standardization of methods will help with some of the demands and requirements for data to assess changes in population demographics. Use of

multivariate modeling and nonlinear equations in assessments may be necessary to clarify trends and identify relevant correlation and covariation.

4.1.1.2 Mortality [H]

Estimates of annual survival and seasonal mortality for different sage-grouse age and sex classes are important for understanding sage-grouse population dynamics. Experimental manipulation of environmental variables, such as habitat or predation, may help assess causal mechanisms for changes in these vital rates. Determining the actual cause of sage-grouse mortality can be extremely difficult, but understanding these patterns would help with population management. Quantitative estimates of mortality are needed in particular life stages (for example, chick mortality during early brood-rearing) to understand how variability in survival affects population growth. Survival during the interval known as "survival to recruitment" appears to be much more variable than adult survival (Moynahan and others, 2006, 2007; Sedinger and others, 2011; Blomberg and others, 2013; Nonne and others, in press).

Population management would benefit from an assessment relating sage-grouse mortality rates and the factors that influence them to the effectiveness of actions taken to reduce them, explicitly considering variation in survival due to age, sex, region, habitat type and condition, and specific management actions. Importantly, estimates of mortality rates would directly inform population models by focusing estimates across the full (multi-season) distribution of the target population and associated habitats. Refer to section 4.1.4, "Mortality Agents and Factors" for additional information about causal mechanisms.

4.1.1.3 Habitat Conditions and Change Agents [H]

Habitat research was not prioritized by the participants at the focus group meeting (table 1) because of the extensive research already done that addresses sage-grouse habitat use (Connelly and others, 2011a). However, ongoing efforts of land-management agencies suggest that additional understanding of ecosystem function and effects of habitat management on sagebrush ecosystems would be useful. A greater understanding of the relationship among disturbance cycles, natural recovery, including restoration and rehabilitation processes, the composition and productivity of vegetation, and site potentials could lead to improvements in habitat conservation and management. For population modeling, accounting of the distribution of quality seasonal habitats and the value of all available habitats, and particularly potential threats to those habitats, could provide a frame of reference for population viability estimates. Importantly, accounting for relations between seasonal habitat conditions and the population response in the same time period would establish connections between habitat quality and survival and discern net-negative population effects from compensatory effects (for example, high early season mortality may be offset by increased survival later in the season). In this context, an understanding of the effects of anthropogenic development and associated land uses on habitat quality, for projections and scenario assessments would support improvements in population models. Development of a spatially explicit population model incorporating current estimates of demography, with appropriate representation of spatial-temporal variation and movements, to evaluate the relative effects of changing land uses on sage-grouse populations would support management planning for all populations of sage-grouse.

4.1.1.3.1 Implications of Priority Areas for Conservation [H]

Because definitions and application of "priority-area" concepts vary by State, questions arise that could be addressed by comparison of strategies and effects on habitat conditions, sage-grouse behavior, and sage-grouse population dynamics. An assessment of the Wyoming Priority Areas indicated that those areas provide better protection for nesting locations than summer or winter locations (Fedy and others, 2012). Similar assessments within each State would help determine the effectiveness of a priority-area approach for conserving habitat necessary for resident sage-grouse populations. If conducted using similar methods, these State-based assessments could be used to identify implementations that have been most successful and help adapt the boundaries of priority areas to better meet conservation objectives.

4.1.1.3.2 Movement Patterns and Connectivity [H]

Connectivity among populations and subpopulations can have major influences on effective population size and is an important habitat consideration (for example, Wisdom and others, 2005a; Aldridge and others, 2008; Bush and others, 2011). Although some assessments of sage-grouse connectivity have been conducted (Knick and Hanser, 2011; Knick and others, 2013), habitat and population connectivity are poorly documented (for example, as evidenced in genetic similarity). Connectivity assessments can help delineate important habitats between (and beyond) priority areas and seasonal habitats, as well as provide information about metapopulation dynamics for use in conservation planning and management efforts. A detailed discussion is in section 4.1.2, "Connectivity."

4.1.1.4 Population Dynamics [M]

4.1.1.4.1 Reproduction [M]

Rate of reproduction is an important population factor. Nesting hens and young broods are sensitive to habitat conditions, disease, and weather events (Moynahan and others, 2006), and they are vulnerable to predation (Taylor and others, 2012). Further, because there is considerable variability in reproduction among sites and years (Taylor and others, 2012), monitoring of reproductive success and research involving questions about causal and correlative mechanisms would support modeling and assessments. For example, determining relationships among the conditions of the hen during the pre-laying period, chicks at hatching, and chick survival would improve models of habitat-hen-reproduction relationships. This relationship may be influenced by genetics or individual fitness, which highlights the utility of research that can discern the difference among multiple determinant factors. The predictive quality of models may improve by incorporating information about relationships among habitat conditions, environmental patterns (for example, weather) and population parameters, particularly pre-laying condition of females, chick survival, and brood-rearing success.

4.1.1.4.2 Juveniles [H]

Differentiation between causes and rates of juvenile sage-grouse mortality from other stages of development is necessary to inform estimations of production; these related variables have important effects on long-term population viability. Estimates developed on range-wide, regional, and local bases would provide information for multi-scale assessments. Developing a better understanding of survival rates, particularly for juveniles under different conditions, also is important for developing effective conservation actions (Blomberg and others, 2012a, 2013; Nonne and others, in press).

4.1.1.4.3 Productivity and Recruitment [M]

Productivity and recruitment are essential components for maintenance of sage-grouse populations at local, regional, and range-wide scales. Accurate and consistent estimation of these attributes, including variability in vital-rate estimates by age of individual, region, habitat conditions, weather, and predation pressure, feed into assessments and conservation planning. Because recruitment at the population level is an aggregate of multiple vital rates, it is important to understand the contributions of each component and life stage and the ecological factors that affect them.

4.1.1.4.4 Implications of Population Cycles [L]

Population cycles have been documented for sage-grouse but remain largely unexplained (Fedy and Doherty, 2011). Cyclical patterns in population numbers may affect interpretation of population trends and minimum viable population estimates. Identification of explanatory factors for cycling at the range-wide, regional, and local population levels would improve the overall understanding of long-term population processes and would help identify individual population drivers in the context of longer-term cycling. This information would improve models and assessments, including those describing population demographics.

4.1.1.4.5 Genetics Applications and Effective Population Size [H]

Application of genetic information would inform conservation decisions addressing population connectivity, isolation, adaptation, plasticity, dynamics, and effective size. The National Conservation Objectives Team report (U.S. Fish and Wildlife Service, 2013), among others, noted the absence of robust range-wide genetic analyses as a potential limitation on long-term sage-grouse conservation efforts. Several ongoing projects have begun to address these issues. Genetic information also would help with population-size estimation, including determination of minimum effective population size

to avoid inbreeding depression, as well as estimation of the population size necessary to balance changes in population genetics caused by drift, mutation, or adaptation.

4.1.2 Connectivity [H]

Understanding connections between populations and subpopulations is important for management of the genetic diversity that helps maintain adaptive and evolutionary potential and avoids inbreeding depression and other genetic disorders. In addition, connectivity of habitats used by each population (or subpopulation) is related to the configuration and condition of seasonal habitat within the home range, as well as lands (habitats or corridors) that are used by sage-grouse during seasonal movements. Thus, the structure and condition of habitats within a population home range is important for the functional use and movement of individuals within that range. Functional connectivity occurs when individuals migrate to a neighboring population and breed. The genetic diversity of the species is maintained by distribution of genetic traits from one population to the other. In addition, small or peripheral subpopulations may act as intermittent population centers (sink populations) for years, but habitat dynamics, climate, land use, and (or) disease may result in these subpopulations acting as refugia, instead of sinks, in the future. Thus, maintaining metapopulation structure and connectivity is understood to be a component of long-term resilience (Gilpin and Hanski, 1991; Hanski, 1994). Research that simultaneously addresses population and habitat connectivity would inform management of habitat configuration and conditions and address circumstances that might influence sage-grouse movement behaviors. The results also would make explicit connections between conservation of genetic diversity, metapopulation structure and maintenance of peripheral subpopulations, and environmental factors that affect habitat use, such as land use and climate.

Connectivity patterns of sage-grouse have been examined with a focus on connections between neighboring breeding habitats (Knick and Hanser, 2011; Knick and others, 2013). These studies have added to our understanding of connectivity patterns, but further investigation of sage-grouse movement patterns, the habitat characteristics that are conducive or restrictive to those movements, and the genetic structure of populations will help inform management practices to improve or maintain connections.

4.1.2.1 Movement Patterns [M]

Sage-grouse are highly mobile and often require large areas to meet their annual habitat needs (Connelly and others, 2011b). Several important questions about sage-grouse movements remain despite the many studies that have been conducted assessing movement patterns using radio-tracking (for example, Berry and Eng, 1985; Aldridge and Brigham, 2002; Holloran and others, 2005; Gregg and others, 2007; Moss and others, 2010) and global-position-system marking (for example, Dzialak and others, 2012; Fedy and others, 2012). Identification of the causes of dispersal and variation in seasonal movement patterns, as well as the habitat conditions suitable for dispersal or seasonal movements, would improve sage-grouse habitat management and provide a context for assessing the importance of areas within the non-breeding portion of the annual cycle. Analysis of the condition of habitats used during dispersal and migration would help determine if these conditions differ from seasonal habitats. In addition, integrated studies that incorporate genetic similarity, habitat connectivity, and details of movement patterns would enhance understanding of metapopulation structure and effective population size.

4.1.2.2 Habitat [H]

Suitable habitat in the matrix between populations and seasonal ranges is necessary to facilitate movement and connectivity. Matrix habitat must provide essential requirements of food and cover, and ultimately facilitate health and connectivity of populations and subpopulations (for example, Wisdom and others, 2005a; Walker and others, 2007; Aldridge and others, 2008; Meinke and others, 2009; Tack and others, 2012). Several fundamental questions about sage-grouse habitat requirements remain despite extensive research involving many populations. Due to differences among populations and habitats in different regions (for example, sage-grouse management zones), it will be important to use data from different regions to answer these questions. These datasets may be available and questions could be addressed using meta-analyses; for questions where data are lacking, regional coordination would improve the effort to develop new data. Important questions that require habitat perspectives will inform research and management through integration with other methods, including requirements and interactions among habitat patch size, size and juxtaposition of seasonal and year-round habitats, habitat linkages important to movements and dispersal, and linkages between small and large populations (for example, Knick and others, 2013). Refer to section 4.2.2, "Connectivity," for additional information regarding habitat connectivity.

4.1.2.3 Barriers and Inhospitable Conditions between Habitats [H]

Information about barriers to sage-grouse movements between populations and between seasonal habitats has come from studies of habitat selection (for example, Bruce and others, 2011; Fedy and others, 2012; Tack and others, 2012). Studies of genetic similarity (Bush and others, 2011) have provided some descriptions of suitable habitat conditions.

However, habitat selection and movement are different processes. An individual sage-grouse may be able to move large distances and through undesirable habitat to reach distant habitats (for example, Fedy and others, 2012; Tack and others, 2012), yet it is not clear that this capability is consistently realized. Critical restrictions on seasonal movements, as when hens move broods from the nest site to late brood-rearing habitat, may affect annual population dynamics and connectivity among populations over long periods of time. An assessment of existing movement data collected using radio or global-position-system tracking technology may identify topographic, vegetative, or anthropogenic features avoided or preferred by sage-grouse while they move between seasonal habitats, migrate long distances, or generally disperse. This type of assessment also may help address questions related to habitat condition and sage-grouse use within designated "General Habitats." Relating sage-grouse use to habitat conditions would help define and maintain migration habitats that support safe movement, forage for stopovers, and otherwise provide connections and intermediate habitats between "priority habitats." Additionally, research could determine the minimum-distance threshold(s) between occupied subpopulations that effectively restrict(s) the movement of sage-grouse.

4.1.2.4 Genetic Evidence [H]

Genetic data can provide evidence for long-term and relatively recent patterns of gene flow following movement and reproductive success of sage-grouse between populations and subpopulations (Benedict and others, 2003; Connelly and others, 2004; Oyler-McCance and others, 2005; Bush and others, 2011). Genetic connectivity is necessary to avoid problems associated with isolation of genetically limited populations, such as inbreeding depression and bottleneck effects. This information is relevant to inform population viability estimates and conduct local population management, for example, conserving habitat connections to avoid isolation of small subpopulations within a management zone.

The basic question, "How are populations and subpopulations connected?" may be usefully asked at local, regional, and range-wide scales. Information about range-wide genetic and habitat connectivity, coupled with knowledge of dispersal dynamics, represented by individual movement patterns through a habitat matrix, would help clarify the causal mechanisms of behavioral responses to habitat conditions and provide meaningful interpretation of landscape connectivity. Related questions include, "What are important landscape features that influence gene-flow movements?" and "How do dispersed individuals and subpopulations relate across the matrix of habitat with other subpopulations?"

Several fundamental questions relate to application of genetic evidence to assess connectivity. These questions consider relations among physical connections between habitats, population movement, and dispersal, genetic change through drift, mutation, or adaptation, likelihood of a bottleneck (or similar, limiting effect) event, and how these factors could affect the long-term survival of the species.

4.1.3 Population Monitoring [H]

A long history of counting sage-grouse at leks has resulted in datasets that can vary by agency, region, population, and in other important ways. Comprehensive evaluation of the strengths and weaknesses of different protocols would help determine which datasets can be combined and effective ways of doing so. This evaluation could help identify standard protocols that could improve estimates of population parameters and reduce variability associated with methodological differences. Many population parameters are of interest to State and Federal wildlife managers because managers often focus on population responses (counts and trends) to manage habitat change and other potential effects. These parameters include female-to-male ratio for annual and regional comparisons; variation in lek attendance by age, time of day, time of year; relationships between lek attendance and peak timing of female nesting and potential relations among environmental drivers and population responses. Further, lek counts can provide an aggregated indication of population dynamics, yet understanding of the mechanistic impacts of habitat and disturbance factors and the life-stage specific responses requires other monitoring approaches. For example, reliance on data from lek counts presents major limitations for understanding mortality of juvenile sage-grouse in relation to extreme weather events or describing variation in nesting or brood-rearing success under changing land uses, such as energy developments. In addition, comparison and calibration of historical (sentinel site) designs with spatially explicit, representative designs would help define sage-grouse abundance and probability of habitat use, which would then increase the usefulness of historical and modern inventory efforts. Furthermore, investigation of the relation between historical survey designs (series of distributed sentinel, lek sites) and spatially explicit designs representing the entire population across seasons would help standardize and modernize population assessments.

4.1.3.1 Lek Counts [H]

Despite the limited portrayal of sage-grouse populations by lek counts (single sex, single season), these counts remain an indispensable component of research and monitoring for conservation. Lack of standard definitions of lek status and standard methods for conducting lek counts affects inferences across large regions and comparisons among populations (Beck and Braun, 1980; Walsh and others, 2004). Importantly, lek counts typically are used as an estimate of total population counts and, over time, an aggregated

indicator of population trends. Estimates may be used for local and regional comparisons, impact studies (of development, for example) and local population management (hunting regulation, for example). Lek counts have not proven to be accurate for understanding regional or population-level differences in important population demographics, such as juvenile survival, because of count biases involving males and season of year. Difficulties with count standardization may arise from lack of endorsement and use of published methods rather than absence of the necessary protocols (Connelly and others, 2004), and differences in project objectives. Standardized methods and analytical approaches would include and explicitly address spatial patterns and temporal variability, while providing parameter estimates for population. Methods that increase statistical power to detect small changes in population size over short time spans would be useful. Beyond these methodological challenges, current uncertainty in lek counts suggests a need to better understand and account for variability in observations over time and the spatial representation of populations (Walsh and others, 2004). Clarity may come from comparison and evaluation of methods to identify the best approaches to effectively survey leks, given the variability in lek counts within a year and between years, to provide relevant numbers for long-term monitoring and annual planning. Similarly, it is important to determine the best methods for surveying lek complexes and (or) best methods for use in Before-After-Control-Impact (BACI) designs to assess development impacts. Advanced modeling and decision-support tools should work to provide projections of future conditions based on consistent, current estimates. For example, definition of the relationship between productivity in one year and the coefficient of variation in lek counts in the subsequent year, with consistent application, would guide both harvest rates and habitat protections.

The location of leks is another monitoring consideration. Not all sage-grouse leks have been located, and the majority of leks are not monitored on an annual basis. A standard approach for searching for new or previously unknown sage-grouse leks could be incorporated into revised sampling designs. Furthermore, not everyone agrees that the relationship between lek counts and true (male and female, multiple age-groups) population numbers is sufficiently understood. The correlation and time lag between the variation in annual sage-grouse productivity and subsequent lek counts affects the precision of population estimates, among other factors. Thus, dependence on lek counts perpetuates uncertainties about the relationship between lek attendance, population numbers, and population dynamics, including the previous years' productivity, rates of inter-lek movements by males, condition of males, climate fluctuations, and importantly, how these variations influence population estimates. For example, previous research demonstrated a significant effect of drought and habitat conditions (only partially correlated) on male attendance at leks (Blomberg and others, 2012a), and although these trends are presumed to mirror trends in the entire population, differentiation of mechanisms and effects on different population segments is not well documented.

Development of a probability-based, spatially balanced sample of breeding males and females would alleviate the limitations associated with non-random population sampling targeting known leks. Garton and others (2011) recommended integrating a wide survey across all habitats, systematic sampling of large leks, and intensive sampling at sentinel leks. This would require the development of a range-wide probabilistic sampling approach, which would lead to an evaluation of limitations associated with extrapolation and trend analysis from the current lek counts. Such an approach also would provide opportunities for more explicit treatment of biases based on location of leks, estimation of population and lek sizes, and development of statistical relationships between lek counts and independent estimates of population size. Ultimately, this could lead to a statistically informed sampling design that reduces the number of lek surveys required for robust population estimates. Further, clarification of the effects of development on lek attendance in comparison to nesting behavior, brood-rearing success, and juvenile survival would be useful for planning and mitigation of industrial activities, and would improve understanding of population dynamics. Thus, it would be beneficial to improve value and cost effectiveness of lek counts by improving connections between counts and population numbers and by accounting for variation in male and female attendance rates and seasonal and daily attendance rates among sampling approaches and among observers, habitats, regions, and topography. A statistically reliable trend-monitoring protocol for inventorying lek attendance of male sage-grouse also would be useful.

4.1.3.2 Demography [M]

As alluded to previously, development and standardization of accepted, common methods for population demographics that result in data necessary to address fundamental questions that cannot be addressed directly by lek counts, are needed. The important relations that need to be addressed surround direct and indirect connections between environmental variables (for example, surface disturbance, range condition, climate and weather events, and (or) predator populations) and population demographics. To estimate the needed parameters and relations, research projects need to use populations, and their entire (multi-season) range, as replicates to monitor the effects of changing conditions within seasonal range conditions on specific components of population demographics. For example, suitable projects would address differences among sage-grouse age distributions, seasonal mortality, nesting, and (or) productivity rates, due to land-use trends, habitat treatment effects, and related spatial patterns, such as habitat loss or fragmentation. Population estimation

methods could be integrated with traditional lek-count surveys to provide the demographic resolution necessary to assess conditions within and between seasons and management units (for example, sage-grouse management zones). Further, the role of population and habitat relationships in other seasons needs to be investigated to assess density-dependent behaviors at different times of the year.

4.1.3.3 Sex Ratios [H]

Sex ratios have been estimated by State agencies with considerable variation in methods and estimates (Connelly and others, 2011b). Establishment of accurate sex ratios with fine spatial and temporal resolution is important for establishing population size when male-focused lek counts are used to estimate population size. Refinement of existing techniques, that is, traditional hunter-harvest-based assessments, and development of new techniques will improve estimates of sage-grouse sex ratios. Resolution of estimated sex ratios should be fine enough to differentiate sex ratios within individual populations. Promising genetic methods have been developed using DNA from fecal pellets (Baumgardt and others, 2013), but these need further testing and refinement for field application. For example, in order to avoid sex-bias, previous knowledge of the distribution of males and females of a population, in a given season, would need to guide the sampling design.

4.1.3.4 Genetics [H]

The most immediate uses for genetic information include an understanding of relationships with demographic patterns and population dynamics of sage-grouse, including sex ratios, male genetic contributions, dispersal, and other parameters that determine effective population size and viability. Although less immediate to population management, an understanding of the genetic connections to morphological, physiological, and behavioral differences among sage-grouse populations across the range will facilitate recognition and conservation of diversity. Continued development of methods and data could provide basic biological knowledge of genetic and phenological adaptations to local conditions, as well as practical applications for conserving genetic variation.

A better understanding of the effects of fragmentation, isolation, and landscape barriers on sage-grouse dispersal and population genetics is needed. Genetic methods, such as micro-differentiation of genetic segments using micro-satellites (Oyler-McCance and St. John, 2010; Bush and others, 2011; Gregory and others, 2012) or amplified polymorphic fragments (AFLP) for detecting short-term differentiation and drift (Veith and Schmitt, 2008) exist, but study designs and specific applications for analysis across landscapes are currently under development.

4.1.3.5 Brood and Juvenile Surveys [M]

Brood and juvenile surveys are not a common practice in the sage-grouse research or management community, yet an evaluation of effectiveness, efficiency, and accuracy of methods (such as Dahlgren and others, 2010b) could determine values of different approaches. Brood surveys provide an alternate method for collecting information that can be applied to the long-term monitoring of sage-grouse populations or to the identification of crucial habitat. Currently (2013), information about juvenile abundance is collected primarily through hunter-harvest surveys. Comparison of several methods for population monitoring would clarify methodological questions, such as how estimates from brood or juvenile surveys compare with lek counts and harvest surveys.

4.1.3.6 Isolated Populations [M]

Monitoring small, isolated sage-grouse populations is complicated by the availability and allocation of time and resources and because physical access to these populations can be difficult. Effective, efficient approaches for determining the demographic attributes of sage-grouse populations in isolated areas are lacking. Use of remotely controlled aircraft (drones) with heat-sensing imaging equipment (remote sensing) has been proposed, but use and methods are not currently established.

4.1.3.7 Pellet Counts [L]

Pellet counts may be an effective, non-invasive, spatially explicit, and temporally explicit method for estimating sage-grouse population numbers and use patterns (Dahlgren and other, 2006; Hanser and others, 2011; Schroeder and Vander Haegen, 2011). To investigate the potential for widespread implementation of pellet counts, core questions first need to be addressed to determine if pellet counts are an effective survey technique for sage-grouse abundance or for obtaining presence and absence information. It is also important to know how pellet-based approaches relate to other techniques.

4.1.3.8 Multiple Scale Relationship and Inference [H]

Development of consistent and representative data for input to models assessing relationships between key demographic parameters and environmental variables (for example, vegetation characteristics, habitat configuration, topography, biogeography, and predator distributions) is important for bridging the gap between large-scale population and habitat monitoring and detailed demographic studies. Even coarse estimates of population trends and cycles, if related to regional patterns of land use, climate, and other anthropogenic factors will support planning and encourage actions to offset any negative trends. The analytical methods

exist (Blomberg and others, 2012a; 2013), but widespread application to sage-grouse has not occurred.

Successful inventory and monitoring methods will likely require implementation of an integrated remote-sensing and field-based sampling design focused on habitat and sage-grouse demographics. The design could be used to simultaneously assess individual populations and local and regional habitat conditions, followed by statistical modeling to develop and define relationships, indices, and interpretations.

4.1.4 Mortality Agents and Factors [H]

Distinguishing causes of mortality, in general, is ranked high as a priority, and State wildlife biologists who are responsible for addressing predation, harvest, and other mortality agents expressed the strongest interest in these topics. Priority issues within this topic are indicative of concerns about interactions of multiple factors, such as interactions between infrastructure and predator distributions, or effects of over-grazing on vegetative cover and nesting success. Seasonal, life-history stage, and habitat-specific mortality effects are used to inform population models and assessments of habitat condition.

4.1.4.1 Sources, Rates, and Influences [M]

Research often focuses on specific sources of mortality independently. Comprehensive assessments of mortality sources that collectively evaluate the effects of predation, insecticides, disease, and other sources of mortality on sage-grouse populations could help identify interactions among mortality sources. Explicit consideration and differentiation of the causes of mortality in different sage-grouse age and sex classes, and the consequences of mortality for population dynamics would inform management. One possibility is to establish correlations between mortality rates and environmental variables. Brood-rearing and juvenile stages are the most important starting point for this research.

4.1.4.2 Predation [M]

4.1.4.2.1 Interactions with Habitat Condition [M]

Management could benefit from better information about predation rates in relation to local and landscape-scale habitat variables. This work could increase the understanding of relationships among habitat structure, population dynamics of sage-grouse, and the predator community within an area (Coates and Delehanty, 2010). It has been suggested that sufficient growth (height, cover, or both) of grasses is required to provide sufficient cover to protect sage-grouse nests and young broods (Sveum and others, 1998a, 1998b; Baxter and others, 2009). Standardization of methods to provide consistent estimates of habitat conditions associated with predation events would help clarify this relation and reduce emphasis on factors that are less important, as well as reduce variability between studies caused by differences in methods and analytical processes. Knowledge about specific predator-habitat associations would help develop future management practices (that is, adaptive management) by identifying features that increase or reduce predation risk, thereby enabling promotion or avoidance of specific practices or patterns. Thus, general knowledge about predation relations is less important than improved understanding of specific relations between infrastructure types and patterns that can be applied in development planning.

4.1.4.2.2 Interactions with Infrastructure [H]

Evaluations of habitat-predation relationships should include the effects of infrastructure, powerlines, roads, and fences on sage-grouse populations. Specifically, to support planning and mitigation, documentation of the incidence and extent of avian predation on sage-grouse nest success, and juvenile and adult survival in areas with and without extensive infrastructure are needed for comparison. Research could document a change in the abundance and distribution of predator populations in relation to changing infrastructure, thereby helping to establish direct and indirect effects-distances (or areas) that can be applied during development planning and in general land-use plans. Application of BACI designs would be useful for demonstrating effects of the infrastructure development. This research also would help elucidate the extent that human-subsidized predators limit individual sage-grouse vital rates and overall population growth. In most cases, these topics have been mentioned, but not assessed.

4.1.4.2.3 Control Effects [M]

Predator control remains a tool for population management in some States; however, the effects of these efforts often are contested and debated (for example, Schroeder and Baydack, 2001; Mezquida and others, 2006). Discrimination of sources and rates of predation by different species would help determine background rates, target concerns, and provide comparative measures of effects. The effectiveness of various predator-control measures, as indicated by positive response in sage-grouse numbers, is an important topic (but see Hagen, 2011b). In addition to measuring the population response of sage-grouse in relation to control efforts, the population response of predator species is needed to recognize compensatory reproductive and behavioral responses that would reduce effectiveness. Further cost-benefit analyses could be used to help inform practicality and effectiveness of predator-control actions.

4.1.4.3 Harvest [M]

Although there is limited evidence of the influence of harvest on sage-grouse population viability, harvest's on-going role and recurrence in management, policy, and public opinion suggests that a collaborative range-wide assessment of harvest effects is needed. This type of assessment could address demographic and population responses to harvest using different bag limits and seasonal timing, issues of additive versus compensatory mortality (when, where, and why mortality occurs), and changes caused by habitat, weather, and management actions.

4.1.5 Population Dynamics [M]

4.1.5.1 Demography [H]

Information about demographic rates as they relate to population growth and effects of habitat loss and fragmentation would inform many aspects of sage-grouse conservation. Accurate estimates of the demographic parameters and population dynamics of sage-grouse, including nesting likelihood, nest success, clutch size, renesting likelihood, renesting success, hatchability, sex ratios, overall productivity, density-dependent effects, male genetic contribution per generation, dispersal, seasonal mortality, and other parameters would be informative to help assess management outcomes, as well as determine effective population size and transitions within populations that affect viability. Further, standardized methods for estimating demographic parameters, coupled with capabilities for resolution across multiple years and seasons, would inform interpopulation comparisons and enable regional assessments. For some traits for which good estimates exists (for example, nest success), a reassessment of the data using modern statistical methods may be needed to overcome limitations caused by outdated methods and potential bias before the results are used to analyze sage-grouse population dynamics.

In the process of developing standardized methods, comparative analysis of existing techniques would inform translation across management units and may facilitate incorporation of historical data. For example, WAFWA's conservation strategy (2006) suggested a sensitivity and elasticity analysis for demographic parameters, which may be used to discern the best (consistent, representative) indicators of population conditions. A better understanding of contributions of demographic components needed for accurate estimation of sage-grouse population growth also would provide a better understanding of the causal factors contributing to long-term declines. But this requires an understanding of how the different parameters associated with productivity compare across regions and management units (for example, sage-grouse management zones or priority areas).

4.1.5.2 Reproduction [M]

Improving estimation of parameters of reproductive success, such as nesting success and survival rates of chicks and juveniles, will improve population models by differentiating timing and causes of reproductive success or failure. Development of spatially explicit empirical models of the connections between environmental patterns, such as habitat productivity, and reproductive parameters, such as pre-nesting condition of hens and nesting success, would provide the ability to map predicted reproductive success. Resulting population models would inform local management of population trends, which rely on timely reporting of population growth rates to establish annual sage-grouse harvest quotas and to develop plans for range management.

4.1.5.3 Productivity and Recruitment [H]

Experts clearly identified the need for accurate measures of recruitment in relation to conditions in nesting and early brood-rearing habitats, particularly for developing management recommendations to support these life stages. Life-history events leading to recruitment, including female nesting tendencies and post-fledging survival, are poorly understood. Although counting males is the most common technique and may be perceived to be the most direct way of attaining estimates of recruitment, a lack of understanding regarding the connections between this estimate and other critical life stages and demographic components make lek counts inadequate and emphasize the value of testing more comprehensive methods (for example, Blomberg and others, 2012a).

Definition and interpretation of factors that regulate chick survival are important for improving monitoring methods to determine chick-survival rates and mortality factors. These improvements in methods and resulting data could aid future comparisons of adult and brood conditions between multiple populations. Assessments examining the role of habitat components and conditions, as well as climate and predation effects, can lead to a better understanding of year-to-year variability in chick-survival and mortality factors. For example, determination of relationships between condition of the hen and the weight of chicks at hatching and relations between brood-rearing habitat condition and chick survival may facilitate adaptations in habitat management.

Slow maturation and low reproductive rates are characteristic of sage-grouse populations (Johnson and Braun, 1999; Beck and others, 2006). Consequently, mortality of juvenile sage-grouse can have a major influence on population productivity and stability. Understanding juvenile survival rates under different conditions is important to develop effective conservation actions because these young birds represent the reproductive potential of the population into the near future.

Understanding the variability in productivity and recruitment in relation to different populations, regions, habitat conditions, weather, predation pressures, and management strategies is important for the development of robust population models. This variability can be explored using various empirical approaches, including spatially explicit models (Guttery and others, 2013). Because recruitment at the population level is an aggregate of multiple vital rates, it is important to understand the contributions of each component and the ecological factors that affect them.

4.1.5.4 Isolated Populations [M]

Small, isolated populations may be as important as large ones to allow for climate-driven habitat shifts and range expansions. They also may be important for maintaining meta-population structure, genetic diversity, and population refugia, but they also may be less accessible than large populations or present other challenges for research and monitoring. Important considerations about meta-population structure, gene-flow, and population viability may surround these populations, and the populations' size and isolation may introduce genetic limitations. These populations also may harbor genetic adaptations allowing them to persist in environmental conditions otherwise not suitable for sage-grouse in the main populations. Size and isolation may present conservation challenges, but these populations also offer valuable opportunities to assess effects of population size on population viability. To help with time and efficiency issues, it would be advantageous to improve the effectiveness of approaches to measure and monitor small, isolated populations with limited access (for example, through remote techniques such as remotely controlled aircraft).

4.1.6 Behavior [M]

Sage-grouse behaviors hold the key to understanding selection and use of habitats, population growth rates, mortality rates, distribution, adaptation, and a multitude of factors that are potentially useful for sage-grouse conservation. Studies that facilitate interpretation of behavior, such as food habits, characteristics and causes of dispersal and migration, seasonal site fidelity, as well as differences in these factors due to sex, age, and region, would inform research and management strategies. Development of behavioral data and integrated interpretations would inform demographic estimates, habitat associations, and population models.

4.1.6.1 Dispersal [H]

Sage-grouse dispersal is a largely unknown process. An understanding of the dispersal mechanism and factors contributing to dispersal rate and distance will improve predictions of population connectivity and habitat use. Understanding natal dispersal and how that process affects spatial structuring of populations also is important, along with knowledge of habitats and features that act as barriers to dispersal and how distance between habitats restricts movement of dispersing sage-grouse. These assessments will be challenging due to the low rates of sage-grouse dispersal. Coordination of multiple studies would achieve sample sizes needed for statistically robust analyses (for example, Fedy and others, 2012).

4.1.6.2 Seasonal Movement Patterns [M]

Sage-grouse populations and subpopulations have different movement patterns and may travel long distances along migration routes between season habitats (Fedy and others, 2012). These movement patterns involve various strategies, including one- or two-stage migratory movements (for example, use of different areas for breeding, summer, and winter), or non-migratory behavior (that is, year-round use of same area; Connelly and others, 2011a). An understanding of how these movement patterns vary by sex, age, region, habitat, landscape, and weather may lead to focused actions to address or avert problems, such as targeted conservation and management projects to improve degraded seasonal habitats or to consider locations of infrastructure.

4.1.6.3 Seasonal Habitat Selection [M]

Sage-grouse use various habitats throughout their annual cycle. Identification of population and season-specific habitats is required for an accurate understanding of each population. Regional analyses help define important relationships between sage-grouse and habitat conditions, but local adaptations, conditions, and history may affect populations differently. Therefore, accurate information about seasonal habitat selection relies on explicit information about connections between habitat associates and local population dynamics.

4.1.7 Food [M]

Interactive effects of climate and land use, including habitat distributions, application of pesticides and herbicides, timing of brood-rearing, vegetation emergence, vegetation flowering, and invertebrate-prey emergence may be interrupted under plausible future scenarios. Current management practices, for example grazing, may influence food availability for broods, but these effects are poorly understood. An understanding of spatial-temporal climate relationships of key forage species, particularly plants, would support assessment of the sensitivity of sage-grouse food availability to potential future conditions. Monitoring insect availability, abundance, and diversity within specific sites to gain an understanding of the species, timing, and locations important to sage-grouse would improve estimation of sage-grouse response to future scenarios.

4.1.8 Adaptation [L]

Questions related to sage-grouse adaptations are low priority, although potential adaptive differences among sage-grouse populations, including adaptation to different and rapidly changing environmental conditions, can affect population viability through population-habitat relationships. Estimates of long-term viability are fraught with inaccuracies due to difficulties projecting future conditions, but fundamentally, the ability of sage-grouse to adapt to environmental variability will orchestrate population responses. Considerable understanding of the genetic variability and phenological plasticity of sage-grouse with respect to environmental patterns would be needed to address adaptation questions.

4.1.9 Translocation [L]

Translocations have been attempted to re-invigorate isolated sage-grouse populations throughout their range with little success (Reese and Connelly, 1997), although, a few efforts have shown promising results when the habitat and management conditions were conducive to reestablishment (Bell and George, 2012; Schroeder and others, 2012). Continued research associated with translocations is a low priority. If the priority changed, protocols would be needed to outline conditions appropriate for translocations, and agreement would be needed about what constitutes success of a project prior to implementation. The genetic adaptability of individuals to local environmental characteristics may influence the effectiveness of augmenting existing populations with birds from different populations (Oyler-McCance and Quinn, 2011). Monitoring to identify populations that are sufficiently robust to allow trapping of birds for transplant would be needed to determine suitable source populations.

4.2 Habitat Management

The complex dynamics of sagebrush ecosystems present a significant challenge to conservation and recovery of systems that support sage-grouse populations (table 4; fig. 3). Conservation of well-functioning sagebrush ecosystems and recovery of degraded ecosystems to functional condition are important to maintain sage-grouse, but variability in regional conditions and land-use histories, coupled with diverse management strategies, suggest that comprehensive and consistent understanding of what constitutes a "well-functioning" sagebrush ecosystem would benefit from further research. Landscape-scale conservation requires an understanding of the habitat patterns and ecosystem processes that support the targeted habitat conditions. However, consistent descriptions of target conditions have been elusive. Defining the targeted habitat conditions involves knowing key relationships between sage-grouse health and habitat use, habitat conditions and configurations, and the processes inherent to well-functioning sagebrush ecosystems. Although some research and development may be directly tied to sage-grouse population responses to habitat conditions and management techniques, research into the basic processes and functions of these semi-arid shrublands is identified as a critical need for management agencies that rely on managing habitat for species conservation.

4.2.1 Maintenance, Rehabilitation, and Restoration [H]

The fundamental ecological processes of functioning ecosystems and management practices that affect these processes, such as grazing, burning, mowing, and others, can determine habitat conditions, and are therefore important considerations for land managers and land-use planners. Ecosystem functions, such as the productivity of vegetation types that provide high-quality food and cover, are directly associated with habitat quality and the near-term health and productivity of wildlife inhabitants, as well as long-term conservation of habitat and species. Managers would benefit from guidance and better understanding of the basic processes and functions that create desirable sage-grouse habitat conditions in order to accurately anticipate the effects of policies and actions. Evaluation of potential pathways and transitions between ecosystem and habitat conditions can lead to pro-active steps to minimize impacts, potentially negating the need for restoration or mitigation. This evaluation also can inform planning and implementation of modifications, such as active and passive restoration projects, control of invasive species, and mitigation of fire effects. The underpinning of successful habitat management is a sound, mechanistic understanding of patterns and processes inherent to a well-functioning sagebrush ecosystem so that these processes can be addressed, and even used, to meet management targets.

Table 4.   Research topics in the habitat management theme and priority designations (low [L], medium [M], or high [H]) based on input from a focus group of representatives from Federal and State agencies.

Topic	Priority designation	Topic No.
Maintenance, Rehabilitation, and Restoration	H	4.2.1
Mitigation	H	4.2.1.1
Effectiveness	H	4.2.1.2
Methods	M	4.2.1.3
Connectivity	H	4.2.2
Priority Areas for Conservation	H	4.2.2.1
Habitat Selection	H	4.2.3
Habitat Quality and Vegetation-Population Linkages	M	4.2.3.1
Genetic Evidence and Tools	H	4.2.3.2
Habitat Condition	H	4.2.4
Understory Vegetation	H	4.2.4.1
Multi-Scale Condition (Monitoring and Research)	H	4.2.4.2
Habitat Quality and Population Response	H	4.2.4.2.1
Variability	M	4.2.4.2.2
Surface Water	L	4.2.4.3
Soils	L	4.2.5
Other Wildlife	L	4.2.6
Restoration and Mitigation Effects	M	4.2.6.1

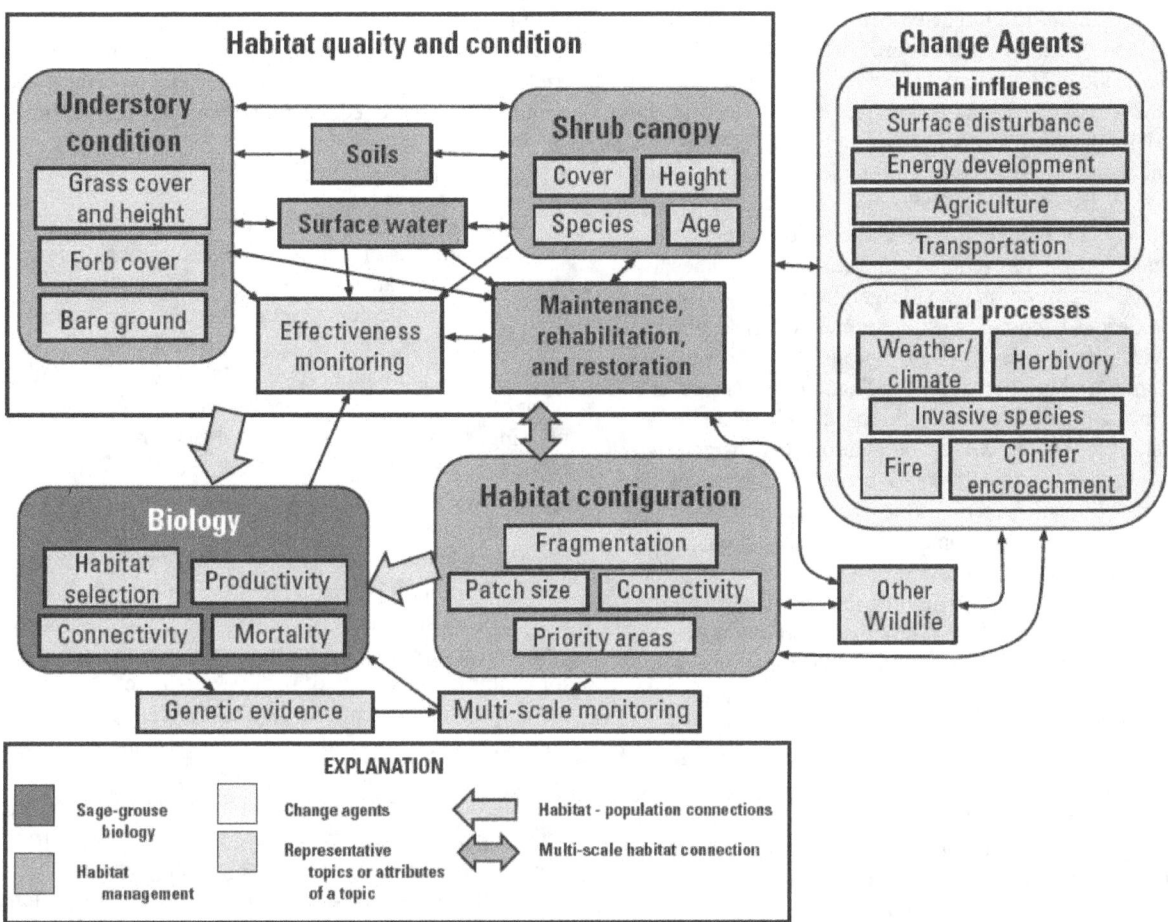

Figure 3.   Connections between research topics and components within the habitat management theme. Boxes represent research topics or important attributes within a theme and arrows provide interactions between themes and topics.

Information about patterns and processes inherent to sagebrush ecosystem function enables successful management, including restoration, of disturbed sagebrush habitats. Restoration science can be likened to aiming at a moving target because the work often occurs in changing combinations of land use, disturbance history, recovery rates, and climate, and where the restored landscape typically is different from the pre-disturbance landscape. Dynamic environmental systems will require perspectives on resistance, resilience, and flexibility of native ecosystems. Analyzing restoration practices in this context can inform site-level choices and activities. In addition, landscape-level prioritization to guide investment and implementation would help determine locations for restoration or mitigation.

Many managers have stated preferences to retain and protect as much intact sagebrush habitat as possible. This goal may not be consistent with all landscapes inhabited by sage-grouse, many of which are currently designated

for multiple use, but it also is unlikely that attempting to protect these communities and landscapes from large, intense disturbances, such as wildfire, will successfully preserve and protect ecosystem function and health (fig. 3). For example, a BLM goal for habitat management for sage-grouse specifies an optimal ratio of 70 to 30 percent, mature to early-phase, sagebrush communities (National Technical Team, 2011). Following from this goal, managers are implementing landscape-scale plans that address the pattern and distribution of habitats. Important perspectives and applications may be developed to assist planning to meet this goal and help balance the distribution of disturbed, restored, and intact sagebrush communities across large regions, such as core areas and priority-habitat designations. It also would be prudent to determine attainability of this specified ratio of habitats and its effectiveness for maintaining sage-grouse populations. Based on recognized requirements of sage-grouse, habitats available now may be recognized as currently suitable (good to excellent condition), impacted with strong potential for recovery using "passive restoration," or impacted with limitations due to condition of vegetation or soils to such an extent that "active restoration" may be required (also see Pyke 2011; Manier and others, 2013).

Understanding of ecosystem functions, resistance and resilience to disturbances (Brooks and Chambers, 2010), and pathways for recovery of desired conditions remain poorly developed and irregularly applied when it comes to restoration, rehabilitation, and mitigation practices. Further, these properties become more confounded when considered across vast, heterogeneous target regions. An improved understanding of relations between "natural" processes, such as fire, drought, and habitat conditions, can elucidate the essential ecological processes and patterns that characterize successful recovery trajectories. Because environmental conditions and disturbance histories vary across the landscape, widespread documentation of applications, along with integrated analysis of restoration practices, succession rates, recovery rates, and environmental covariates would to provide an understanding of processes and influences to guide successful implementation. Implementation of successful management practices at landscape scales, while prioritizing the protection of existing sagebrush communities and rehabilitation of degraded areas by minimizing sagebrush removal and maximizing sagebrush recovery, could benefit sage-grouse populations. Notably, the efficacy of "priority-area" concepts for protecting all or a portion of seasonal habitats and stabilizing population trends within designated priority areas was identified as important research need.

4.2.1.1 Mitigation [H]

Mitigation planning is closely aligned with efforts to prioritize the landscape for conservation because a fundamental component of the planning process is identification of places that have value but need protection. Agreement on the currency for mitigation, which could be defined in various ways, is important. Currency could be economically based in the form of funds expended or management based and measured as acres treated. A sage-grouse-oriented currency, such as number of sage-grouse per unit area, most directly links to sage-grouse conservation. Evaluation and prioritization of target landscapes for mitigation also is important, and includes definition of anticipated benefits to wildlife and evaluation of the suitability of mitigation as a means of providing usable and used habitats. Investing in restoration to repair or otherwise manipulate patterns of use by wildlife assumes that project success will include renewed animal activity. This may not be a valid assumption. Thus, at regional scales, research that examines efficacy of recommended mitigation could help avoid, minimize, or reduce the effects of surface-disturbing activities on-site, or replace or enhance suitable habitat off-site. An understanding of the effectiveness of rehabilitating or restoring

sage-grouse habitat would help determine if off-site mitigation is a viable option, because if restoration of sites to conditions useful for sage-grouse is not possible (for near-term benefits) then mitigation procedures might warrant reevaluation.

4.2.1.2 Effectiveness [H]

Evaluation of the effects of historical and modern habitat treatments on current habitat conditions, use by sage-grouse, or value to sage-grouse may be key for improving management practices and refining treatment techniques. Effects of management actions and related disturbances, including prescribed fires, wildfires, invasive-plant control, juniper (*Juniperus* spp.) removal, mowing, plowing, seeding, chaining, and other forms of sagebrush reduction are best evaluated based on current conditions (to describe need) and value of previous treatments (to indicate probability of success). These evaluations are most effective when they include different habitat treatments, consider different potential roles and values of treatments, and include assessment of the response by sage-grouse. In addition, comparison of different methods and applications, for example, the effects of passive versus active restoration techniques, could demonstrate and differentiate restoration approaches for degraded sagebrush range and help inform state-transition models, which then could guide future work. Further, discrepancies in restoration condition, restoration rate, and "release" of sites (from lease bonds and use restrictions) when compared to historical, pre-existing, or other defined target conditions need documentation. Without a systematic and thorough approach to these issues, effects of historical treatments, effects of development and reclamation, and effects of planned activities cannot be accurately assessed. Furthermore, establishment of common sampling, methods, protocols, and metrics for monitoring effectiveness of restoration treatments at local, regional, and range-wide scales would benefit application and interpretation of the role and habitat values of treatments.

4.2.1.3 Methods [M]

Development and refinement of existing methods for restoration of habitat are ongoing tasks for managers. Empirical data will help distinguish effects and effectiveness of different restoration methods in different environmental circumstances. A comprehensive assessment of past management practices and policies can aid in the identification of those that have had long-term success maintaining or recovering the sagebrush community and sage-grouse habitats. For example, when assessing practices and policies successful in recovering sagebrush habitat, it is important to establish the time frame necessary to achieve an objective. Another consideration is what techniques are effective for improving herbaceous diversity and density, including the forb component, in previously degraded or restored habitats.

4.2.2 Connectivity [H]

Physical proximity, juxtaposition, and suitability of habitats for safe movements (that is, structural connectivity) are important considerations for sage-grouse management because sage-grouse are a widely ranging species and their habitat is fragmented by land uses and infrastructure. Maintenance or restoration of habitat conditions that support safe movements of individuals between seasonal habitats are vital for population viability. Further, connectivity among populations and subpopulations is important for maintaining genetic diversity. Improvements in habitat connectivity are one way that managers can directly affect the ability of individuals to move and populations to mix. Knowledge of habitat components that facilitate or limit connectivity between populations and among seasonal habitats within populations is important, as this will inform what constitutes a connected landscape.

Connectivity has been identified as an important aspect of conservation for research and development. Although connections between landscape configuration and wildlife health and vitality exist (Saunders and others, 1991; Ryan and others, 1998; Bennett, 1999), tests of the ability of restored habitats to support sage-grouse population connectivity through dispersal and migration are lacking. It is not clear what value corridors and habitat islands have for connecting sage-grouse populations but maintaining connectivity between seasonal habitats has been deemed essential (Fedy and others, 2012). Importantly, how habitat characteristics, environmental patterns, and related circumstances interact to affect sage-grouse behavior (use of those habitats) may be useful for predicting valuable locations for restoration. Identification and evaluation of connectivity linkages to document importance to sage-grouse movements and dispersal are common local and regional priorities, and connections between small populations and large populations are of particular concern. Once identified, prioritization of "intact" existing habitats, large areas, connected networks and connective habitats and relationships to population connectivity, dispersal, and migration can inform conservation and mitigation planning.

Connectivity analyses examine connections between components and can be used to examine relationships between sage-grouse and their habitat. Important habitat components include landscape parameters that represent patterns of use (for example, minimum sufficient habitat patch size) and avoidance (for example, surface disturbance). Knowledge of these components can help guide assessments of migration, other forms of movement, and the habitats involved (Barrows, 2011; Fedy and others, 2012; Tack and others, 2012). For any given population, identification of the areas needed to meet minimum habitat requirements (that is, conditions suitable for persistence of the species) to support year-round habitat needs also can be used to assess probable and plausible travel routes between those habitats (Knick and others, 2013).

Many of the questions in this topic relating to relative importance of habitat patterns, conditions, and connections may be addressed by combining telemetry data with land-cover and land-use data to characterize selection of habitats for movement, lingering, and residence. Land managers, faced with orchestrating the trade-offs among multiple mandated uses, are interested in understanding the difference or relative importance between improved connectivity of habitats or subpopulations and increased habitat area. A common consideration is whether configuration is more important that total area. Other considerations include requirements for the minimum habitat area, the maximum distance of travel, the effects of habitat condition within migration habitats, the extent that populations were connected before recent land-use changes, and site fidelity or plasticity in habitat selection by sage-grouse. Addressing these questions requires an empirical model-driven approach using spatially explicit behavioral, genetic, and habitat-condition data for comparison of different populations and scenarios.

4.2.2.1 Priority Areas for Conservation [H]

Additional information is desired about the condition and importance of habitat features necessary for maintaining connectivity within and among populations protected by priority-area designations and those outside of those delineations. The condition and configuration of habitat patches within priority areas, and the relation of these habitats with configurations and conditions outside the priority (protected) areas are essential attributes of these areas that need further study. Priority area designations represent, from one perspective, very-large-scale adaptive management experiments, and understanding the details of conditions that will affect connectivity between these areas and those effects on local sage-grouse populations and regional conservation efforts is essential for informing revisions of designations and future implementations.

4.2.3 Habitat Selection [H]

Sage-grouse habitat selection has been studied throughout the range of the species on a local and regional scale. Although many consistencies can be seen across the range, there are sufficient differences in vegetation characteristics and available habitats to make inferences across populations difficult. Recent studies have addressed the underlying consistencies in animal-habitat relationships using modeling approaches to identify habitat characteristics that vary least within breeding habitats, and thus, infer a set of essential requirements (Knick and others, 2013). Other studies have used empirical models to describe habitat selection on a seasonal basis in Wyoming (Fedy and others, written comm.). These empirical models, informed by data from across the range, would improve understanding of habitat selection by sage-grouse and help with the prioritization and management of sage-grouse habitats. Further, condition of selected patches is best assessed in conjunction with pattern and juxtaposition of available habitat patches, along with local conditions, to improve understanding of interactions between multiple scales of selection.

4.2.3.1 Habitat Quality and Vegetation-Population Linkages [M]

Models linking sage-grouse demographic processes to the underlying causal mechanisms are the next step beyond empirical models of habitat-selection patterns. These include models of bird health in relation to habitat parameters, including habitat condition, quality, and configuration. Uncertainties to be addressed include (1) which sagebrush taxa are being used by various sage-grouse populations and for what purposes during each season, (2) how do habitat patterns relate to nutritional quality and bird body condition, and (3) what are the differences in habitat selection associated with sex, age, season, management, region, weather, breeding success, and survival. For example, in Idaho, recent results indicate sage-grouse use plant chemistry as a criteria for habitat selection (Frye and other, 2013). These analyses could help identify causal relationships, and if conducted in a multiple-scale, spatially explicit context, these analyses may better inform planning and management than contexts that simply address correlative patterns. Linking the structure and spatial organization of vegetation communities and demographic rates can inform planning and targeted habitat management.

4.2.3.2 Genetic Evidence and Tools [H]

Advances in the identification of fine-scale genetic markers will improve studies of genetic links to behavior and sage-grouse population responses to management actions. This emerging field of study has begun to investigate sage-grouse related questions and is appropriate for multiple lines of inquiry. It could be used to study possible linkages between genetic traits and dispersal patterns or between genetic traits and different degrees of habitat fragmentation or other management actions. Genetic studies also could reveal the relative influence of different habitat features on genetic differences, the conditions and patterns that encourage or inhibit gene flow, and any possible seasonal or regional variations in these conditions and patterns.

4.2.4 Habitat Condition [H]

4.2.4.1 Understory Vegetation [H]

Although an abundance of sagebrush is consistently recognized as an important, defining characteristic of healthy sage-grouse habitats, a healthy herbaceous understory also is an important component of quality habitat, and this resource base is both dynamic and readily affected by natural and anthropogenic processes. The importance of herbaceous vegetation to habitat selection and nesting success has been documented (Watters and others, 2002; Holloran and others, 2005; Beck and others, 2009; Hess and Beck, 2012a; Kirol and others, 2012), but regional assessments and specific guidance for providing herbaceous vegetation as habitat are typically lacking. Limitations of remote sensing using satellite or airborne platforms have led to regional habitat-condition assessments focused on measures of habitat condition or ecological communities' classifications that are coarse scale and based on overstory vegetation (for example, U.S. Geological Survey, 2008, 2011; Homer and others, 2009). As a result, few studies have examined the regional effects of the condition of understory vegetation on sage-grouse habitat quality. In addition to assessments of understory conditions, additional work to determine the causes of suppressed herbaceous understory (for example, soil condition, grazing management, and drought) would help identify and implement appropriate methods for improving understory conditions.

4.2.4.2 Multi-Scale Condition (Monitoring and Research) [H]

Long-term assessments of habitat condition at multiple spatial scales occur with coordination of survey efforts across the sage-grouse range. Landscape-scale habitat monitoring currently (2013) is being conducted by programs such as the NRCS, National Rangeland Inventory project (NRI) and BLM, Assessment, Inventory and Monitoring project (AIM). Evaluation of monitoring methods used in these programs would help determine their applicability for monitoring sage-grouse habitat conditions at multiple spatial scales. Generally, these programs may be used to form the core methods for integrated, range-wide monitoring, but there are important questions about applicability. Clearly, the way these monitoring schemes are implemented across multiple jurisdictions and multiple time periods affects their usefulness for assessing range-wide habitat conditions. Linking multiple spatial scales and traversing the spectrum of existing habitat-monitoring schemes can help identify the best sampling strategies. Aggregation of site-level estimates up to regional and range-wide scales also may be desirable for integrated, multiple-scale evaluations. Ultimately, methods and sampling designs necessary to link habitat conditions and land-use patterns to sage-grouse behavior, distributions, and demographics could inform monitoring designs and help bridge the gap between questions addressed by short- and long-term research.

Development of a universal method and process for evaluating and monitoring short- and long-term changes in habitat conditions at a range-wide scale is needed. Research could help identify methods and data sources for repeatable mapping of land cover, species-specific canopy cover of sagebrush, age distribution of sagebrush, and herbaceous understory of sagebrush habitats. A range-wide evaluation of the herbaceous understory component of sage-grouse habitats based on remote sensing has been limited by the availability of suitable technology and image products, but approaches that link remote sensing with field assessment could support range-wide measurement and monitoring of landscape-scale habitat characteristics with good accuracy for most components (Homer and others, 2012). The repeatability of the methods is vital, and rapid updates of products are necessary to detect short-term changes in habitat condition.

4.2.4.2.1 Habitat Quality and Population Response [H]

As is the case for other topics, useful interpretation of habitat conditions can be gained from simultaneous assessment of habitat conditions and wildlife-use patterns. Therefore, research that elucidates connections, particularly those that are causal between habitat quality and effects on sage-grouse populations, including use, movements, individual health and population demographics, will potentially improve habitat conservation and management. Further, use and value for sage-grouse may vary in different seasons and life stages; therefore, projects that address details of use and selection by grouse are most appropriate. Specifically, in order to facilitate habitat management, relationships between sage-grouse population indicators (for example, seasonal use, nesting success, and mortality rates) and habitat parameters including sagebrush-canopy height and cover, forb and grass height, overall plant diversity and abundance, and nutritional quality warrant further study.

4.2.4.2.2 Variability [M]

Sage-grouse habitats across the species' range vary by species composition of overstory and understory vegetation, quality (for example, nutritional value), and configuration (for example, patch size). It is important to improve the understanding of the relationships between the edaphic, topographic, climatic, and disturbance gradients and habitat conditions suitable for sage-grouse. The sage-grouse management guidelines (Connelly and others, 2000a) outline habitat conditions for sage-grouse seasonal habitats with a caveat for regional variation in achievable results. Identification of factors limiting portions of the landscape from achieving local or regional conditions suitable for sage-grouse seasonal use should help inform future management actions and expectations of success.

4.2.4.3 Surface Water [L]

The importance of surface and subsurface water flow for sage-grouse and sagebrush habitats is poorly understood. Research could assess the effects of water developments that range in size from small impoundments to large diversions. A related consideration is how the condition of the surrounding sagebrush ecosystem is changed by alteration of flow regimes when diversions are present. In addition, research to address the importance of free water for sage-grouse is relevant (see Connelly and others, 2011a). These issues will have additional implications if climate change alters the distribution of water.

4.2.5 Soils [L]

Soils are important predictors of vegetation condition and potential. Questions related to soils and the indirect relationships between soils to sage-grouse were a low research priority during the review process. However, the predictive capabilities associated with soil and vegetation relationships (Schlaepfer and others, 2012) can be useful for management. The development of consistent, range-wide soils data and models that describe how changes occur in response to stress and disturbance are valuable tools for managing across landscapes. The Soil Survey Geographic Database (SSURGO) provides a consistent framework for these data and models, but information gaps exist for key areas within the sage-grouse's range. In general, it is relevant to connect ecological site descriptions and state and transition models to specific types of restoration and dynamic ecosystem processes. In addition, these models could incorporate links to sage-grouse habitat requirements and habitat quality at the scale of regions or sage-grouse populations.

4.2.6 Other Wildlife [L]

The sagebrush ecosystem provides habitat for more than 350 wildlife species (Wisdom and others, 2005b). Conservation and management for sage-grouse may influence these species. For example, studies indicate that sage-grouse may act as an umbrella species for passerine birds (Hanser and Knick, 2011), which means that they might benefit from conservation measures focused on sage-grouse. However, links between sage-grouse and passerine population dynamics or the occurrence of other taxonomic groups (Rowland and others, 2006) remain questions for future research. Some advantages and disadvantages likely will be realized by various species in areas where sage-grouse are the focus of management, including the ongoing designation of areas as priority and general habitat.

4.2.6.1 Restoration and Mitigation Effects [M]

Various game species, such as mule deer (*Odocoileus hemionus*), are managed within the range of sage-grouse. Management decisions could be informed by information about the potential for conflicts between the desired conditions created by actions focused on other species and habitat management goals for sage-grouse, including how sage-grouse and other species are affected by management for the other, and whether the effects vary by region, habitat, or other factors.

4.3 Change Agents

The term "change agents" is often used in land-use planning and evaluation to identify those factors that directly or indirectly affect wildlife populations or habitat conditions. These agents may include positive or negative effects to sage-grouse, sagebrush habitat conditions, or both (table 5; fig. 4). Similarly, some change agents clearly are the result of human activities, others are natural processes, and many are a combination of the two because natural processes have been altered by human activities. Some change agents have unique mechanisms of influence on populations or habitat, but they are typically linked in time and space, and therefore, they may act synergistically to influence ecosystem processes (Leu and others, 2008). For example, fire is a natural process, but human influences on fire regimes also are an important component of contemporary habitat patterns and functions. In addition, the interaction between fire and invasion by non-native annual plants works synergistically to degrade sagebrush habitat conditions (D'Antonio and Vitousek, 1992). Although an understanding of individual mechanisms is important, knowledge of cumulative effects is essential for developing best-management practices that provide long-term conservation of sage-grouse populations and functioning

sagebrush. This direction of research is consistent with previously stated strategic priorities, which indicated that large-scale, integrated assessments that consider interactions in space and time are needed to develop understanding at scales appropriate for habitat management.

4.3.1 Anthropogenic Influences

Human actions and infrastructure influence sage-grouse populations and habitat by changing the condition and distribution of suitable locations for breeding, nesting, brood rearing, and wintering. The negative impact of these influences was recognized by the FWS in the most recent sage-grouse listing decision (U.S. Department of the Interior, 2010). Increased knowledge of the relationships between sage-grouse habitat use, demography, and anthropogenic activities, as well as mechanisms of human activity that degrade or improve conditions, could advance management of sagebrush landscapes by informing future development actions, averting negative effects of proposed actions, and meeting obligations for multiple use of public lands. In this Research Strategy, the accumulation of anthropogenic influences and their potential interactions is summarized in section 4.3.1.1, "Surface Disturbance."

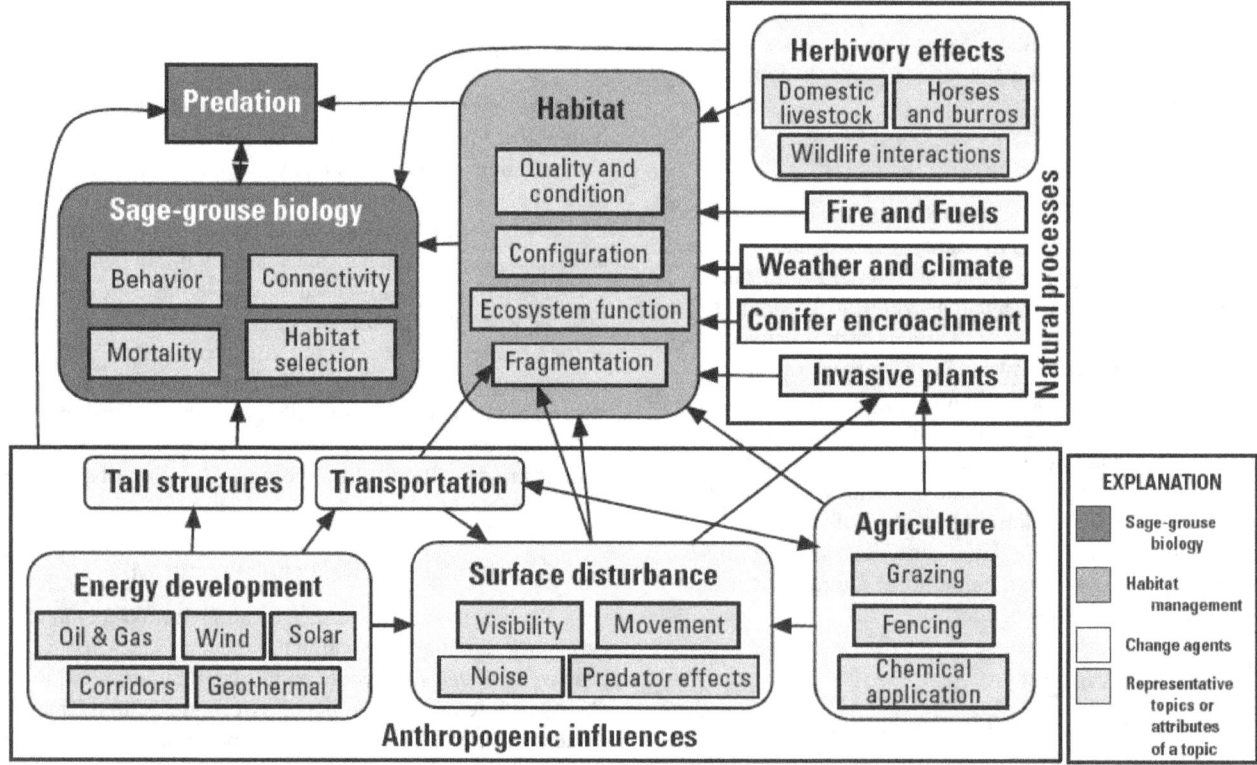

Figure 4. Conceptual model of change agents that outlines important relationships and influences of these factors on sage-grouse and sagebrush habitats. Boxes represent research topics or representative attributes within a theme and arrows provide interactions between themes and topics.

Table 5.   Research topics in the change agents theme and priority designations (low [L], medium [M], or high [H]) based on input from a focus group of representatives from Federal and State agencies.

Topic	Priority designation	Topic No.
Anthropogenic Influences		4.3.1
Surface Disturbance	H	4.3.1.1
Noise	H	4.3.1.1.1
Predator Effects	H	4.3.1.1.2
Movement	M	4.3.1.1.3
Visibility	M	4.3.1.1.4
Energy Development	H	4.3.1.2
Oil and Gas	H	4.3.1.2.1
Wind	H	4.3.1.2.2
Corridors	H	4.3.1.2.3
Geothermal	L	4.3.1.2.4
Solar	L	4.3.1.2.5
Tall Structures	H	4.3.1.3
Agriculture	M	4.3.1.4
Transportation	L	4.3.1.5
Natural Processes		4.3.2
Conifer Encroachment	H	4.3.2.1
Invasive Plants	H	4.3.2.2
Habitat Condition and Ecosystem Function	H	4.3.2.2.1
Restoration	H	4.3.2.2.2
Fire And Fuels	H	4.3.2.3
Landscape Dynamics and Connectivity	H	4.3.2.3.1
Habitat Condition/Effects and Recovery	H	4.3.2.3.2
Planning and Control Methods	H	4.3.2.3.3
Restoration and Rehabilitation	H	4.3.2.3.4
Vulnerability and Prioritization	H	4.3.2.3.5
Population Response	M	4.3.2.3.6
Interactions with Climate, Grazing and Other Land Uses	M	4.3.2.3.7
Herbivory Effects	H	4.3.2.4
Domestic Grazing	H	4.3.2.4.1
Practices, BMPs, Systems	H	4.3.2.4.1.1
Monitoring Effects and Conditions	M	4.3.2.4.1.2
Horses and Burros	M	4.3.2.4.2
Wild Herbivores and Herbivore Interactions	L	4.3.2.4.3
Disease	M	4.3.2.5
West Nile Virus	M	4.3.2.5.1
Background Level of Disease and Implications for Population Cycling	L	4.3.2.5.2
Weather and Climate	M	4.3.2.6
Implications for Priority Areas	H	4.3.2.6.1
Demographics	M	4.3.2.6.2
Cycles and trends	L	4.3.2.6.3

4.3.1.1 Surface Disturbance [H]

Surface disturbance is a combination of the vast array of anthropogenic activities that alter or remove natural sagebrush communities, which means that most research addressing surface disturbance automatically encompasses a wide range of anthropogenic activities. Others have established that sage-grouse are sensitive to surface disturbance in occupied habitats (Doherty and others, 2008; Johnson and others, 2011; Naugle and others, 2011a). Additional research could distinguish the effects of different types of disturbances and related activities on sage-grouse, for example, effects of development of oil, gas, and wind resources. Important topics include effects of disturbances in different seasons and under various habitat conditions, such as sagebrush type, topography, fire history, and vegetation treatments. These studies could focus on identification of thresholds beyond which effects on sage-grouse behavior or population response(s) are minimized. Thresholds could be evaluated based on distance from the location of the surface disturbance, intensity of the disturbance, and similar criteria describing habitat patterns. For example, the Sage-Grouse National Technical Team (2011) outlined levels of acceptable surface disturbance within priority areas for conservation as discrete anthropogenic disturbances covering less than 3 percent of the total sage-grouse habitat regardless of ownership. Additional research could evaluate the effectiveness of this standard with measures of sage-grouse response.

The ecological influence of anthropogenic activity can extend beyond its own physical footprint (Leu and others, 2008) and may take the form of increased noise, changes in predation, and behavioral changes, such as altered movement patterns and habitat use. Additionally, outcomes of widely dispersed disturbances, such as invasive plants and dust, on habitat conditions are poorly understood. Understanding immediate and cumulative effects of surface disturbances could improve siting of future infrastructure developments, mitigate deleterious effects of existing developments, and present new options for management of landscapes for multiple uses. In addition, physical habitat loss may decrease population and habitat connectivity when losses are substantial enough to create functional barriers.

4.3.1.1.1 Noise [H]

Noise has the potential to influence sage-grouse behavior, including habitat use, movement patterns, and breeding activities (Blickley and Patricelli, 2012), with associated implications for population characteristics. Sage-grouse have been shown to have elevated corticosteroid levels when subjected to increased noise (Blickley and others, 2012). Additional field studies could help distinguish the effects of noise and elevated stress hormones on population vital rates, demographic trends, and seasonal patterns of space use and sensitivity to noise. Considerations include responses to intermittent versus continuous noise and the potential influence of wind direction and topography on noise effects.

4.3.1.1.2 Predator effects [H]

Construction of utility corridors, communication towers, wind turbines, and other infrastructure may influence distributions and hunting effectiveness of predators. If construction does occur, it could affect sage-grouse mortality, and through indirect means, affect sage-grouse habitat use. The magnitude and direction of such effects may vary depending on environmental factors and construction designs that affect use of structures by aerial predators as perches for hunting or as nesting platforms. Understanding links between infrastructure, predator population size, and subsequent changes in predation rates on sage-grouse are important for the development of effective management and mitigation strategies.

4.3.1.1.3 Movement [M]

Surface disturbance changes the quantity and configuration of habitats, which may influence movement patterns (Lyon and Anderson, 2003; Holloran and others, 2010). Additional research would reveal the effects of surface disturbance on sage-grouse movement patterns and clarify the potential for various disturbance types and configurations to serve as barriers to movement and connectivity (Refer to sections 4.1.2, "Sage-Grouse Biology, Connectivity" and 4.2.2, Habitat Management, Connectivity").

4.3.1.1.4 Visibility [M]

Infrastructure often has different elevations and visibility profiles than natural environmental features. These differences may influence the perceived risk from these structures and associated population level effects. Research could assess the influence that visibility has on sage-grouse behavior and identify possibilities for adjustments in construction design and materials to mitigate potential influences of infrastructure development on sage-grouse behaviors and habitat use.

4.3.1.2 Energy Development [H]

A host of potential effects, measured and assumed, have been associated with industrial development of public lands for energy resource extraction (Walker and others, 2007; Tack, 2009; Naugle and others, 2011a, 2011b). The longest, most variable, and most contested developments in these landscapes are oil and natural-gas wells (Braun and others, 2002; Holloran, 2005; Walker and others, 2007; Copeland and others, 2009; Harju and others, 2010; Hess and Beck, 2012b). The presence of these wells is accompanied by networks of roads, pipelines, power lines, pumping stations, and consolidation facilities. Various changes have been associated with oil and gas developments (Connelly and others, 2004; Taylor and others, 2012), including new traffic associated with daily maintenance of equipment (Blickley and others, 2012), intense activities and noise during drilling, fragmentation, dust and weeds associated with road networks (Connelly and others, 2004; Bergquist and others, 2007), and new opportunities for predators. New developments and technologies, such as wind-turbine arrays, solar arrays, geothermal facilities, coal-bed methane wells, and hydraulic-fracturing installations, are assumed to have similar effects in cases where infrastructure and activity levels are similar. This assumption is largely untested, and further, activities and infrastructure associated with these developments are not identical (for example see, LeBeau, 2012). In addition, different locations have unique attributes that may affect sage-grouse or habitats in unique ways and may require specific research and management attention.

Improved understanding of disturbance intensity and population response is important, and efforts to gain essential knowledge would include testing of implemented and proposed stipulations (for example, buffer distances and development density). Identification of thresholds and accounting for spatial variability across regions also are important topics.

4.3.1.2.1 Oil and Gas [H]

Most research documenting the response of sage-grouse populations to intensive land use has focused on roads, wells, pads, pipelines, and related infrastructure for oil and gas extraction (Walker, 2007; Doherty and others, 2008; Carpenter and others, 2010; Harju and others, 2010; Holloran and others, 2010; Doherty and others, 2011; Hess and Beck, 2012b). Some aspects of land-use intensity, fragmentation, noise, and buffer distances can be investigated in association with oil and gas development, but most sage-grouse biologists are interested in identifying landscape priorities in order to conduct conservation and mitigation, improve restoration and rehabilitation practices, and potentially influence siting of new developments to minimize impacts and reduce the need for mitigation. Common goals are to decrease the presence or influence of invasive plants, increase native vegetation, particularly sagebrush cover, and increase sage-grouse use of restored areas. Pad reclamation activities provide an opportunity for research and development of new methods for restoration of functional sagebrush communities.

4.3.1.2.2 Wind [H]

Wind-energy development is a relatively new change agent within the range of the greater sage-grouse. The pace of change is rapid as agencies and industry strive to meet the U.S. Department of Energy's goal to have wind power supply 20 percent of the U.S. power generation by 2030 (U.S. Department of Energy, 2008). Protocols have been developed by groups, such as the National Wind Coordinating Council, to study the effects of wind-energy developments. The primary focus has been on the net reaction of sage-grouse populations rather than mechanistic assessment of specific, direct or indirect, effects through time. Birds avoid the vicinity of wind facilities, but the causal mechanisms are poorly understood.

An understanding of what sage-grouse avoid, thereby causing displacement, will inform decisions regarding siting and permitting of facilities that reduce impact to sage-grouse populations. Wind turbines are tall structures and may be perceived as potential raptor perches. They also may influence individual bird movements because of the noise, motions, or human activity associated with wind production. Long-term effects may manifest themselves through changes in the survival rate in areas surrounding wind facilities, and these changes may only occur during certain life stages or seasons. The potential for lag effects in population responses confounds research into these causal mechanisms. Therefore, addressing these issues may require long-term monitoring. A complete understanding is not essential for immediate planning.

4.3.1.2.3 Corridors [H]

Different types of linear features, referred to as corridors, are associated with most of the dispersed energy activities in sage-grouse habitats. These are placed to gather and distribute power and products. An understanding of the effects of existing and proposed energy corridors and associated facilities on sage-grouse and sagebrush habitats would include a combination of observational studies, field sampling, and scenario modeling to address current and potential influences. Effects to be considered include fragmentation, invasive species, noise, and predation. Research designed to include experimental controls and data collection before and after development occurs would greatly enhance understanding and the extrapolation of results.

4.3.1.2.4 Geothermal [L]

The mechanisms of disturbance caused by geothermal-power generation are similar to oil and gas development (discussed in section 4.3.1.2.1, "Oil and Gas") and can be attributed primarily to a proliferation of roads and pipelines. However, due to a limited number of operating facilities in sage-grouse habitats, little direct information about the relationship between sage-grouse and geothermal-power facilities exists. Potential for increased geothermal energy production in the sage-grouse range is located primarily in Nevada and southern Oregon (Knick and others, 2011). In these areas, this topic may be a locally important issue.

4.3.1.2.5 Solar [L]

Large-scale solar developments are not common within the sage-grouse's range at this time, but improvements in technology and incentives to increase use of renewable energy resources may change this situation. Studies addressing the potential effects of solar developments on sage-grouse could help minimize negative effects if these developments begin to occur. Topics of study could include effects on vegetation, predation, and changes in availability of water supplies.

4.3.1.3 Tall Structures [H]

The addition of tall structures to sagebrush shrublands, which lack naturally occurring vertical structures, and the potential effect of those structures on sage-grouse populations has previously been identified as an important research need. The proposed construction of several large transmission lines prompted the development of protocols for assessing the impact of tall structures (Utah Wildlife in Need, 2011). These protocols have not been implemented, and the questions they address remain unanswered. Important considerations are possible avoidance by sage-grouse, reasons for avoidance, changes in predation rates, and contributions to habitat fragmentation. Each of these considerations can pertain to a variety of tall structures, for example, power lines, cell towers, and wind turbines. Established protocols could be helpful for addressing comparable considerations associated with these different types of structures. Assessment of impacts caused by the spectrum of tall structures may help inform future design and siting considerations for these types of development projects and help identify modifications that could be made to existing structures to mitigate negative effects.

4.3.1.4 Agriculture [M]

Agricultural conversion has been associated with decreasing lek trends (Johnson and others, 2011) and increased risk of sage-grouse extirpation (Aldridge and others, 2008; Wisdom and others, 2011). In addition, pesticide application within agricultural fields can have toxic effects on sage-grouse (Blus and others, 1989). Programs established under the national Farm Bill, particularly the Conservation Reserve Program (CRP), have had positive influences on sage-grouse populations (Schroeder and others, 2011). Several factors warrant further study to maximize the benefits of such programs. Considerations include where lands are most effectively set aside under the Conservation Reserve Program to benefit sage-grouse; the appropriate size, configuration, and juxtaposition of these set-aside lands; and the effectiveness of various habitat modifications on set-aside lands. Questions could be answered using relative use, reproductive success, movement patterns and other metrics to differentiate between source habitats versus sink habitats, differentiate use patterns, and differentiate costs and benefits across seasons of use.

4.3.1.5 Transportation [L]

Although not singled out as a research priority, roads are a large source of surface disturbance within many types of landscapes across the sage-grouse range. Accurate, high-resolution road datasets, with all roads categorized by type (for example, unofficial, off-highway vehicle, unpaved, paved) throughout the sage-grouse range would be useful in efforts to account for the effects of this surface disturbance. These datasets also could provide the information needed to further differentiate the effects of road size, traffic levels, and nature of disturbance (noise, dust, visibility, and associated

infrastructure) to sage-grouse. Roads are an important conduit for the introduction and spread of invasive species (see section 4.3.2.2, "Invasive Plants"), and fine-resolution transportation maps may help manage issues associated with invasive species.

4.3.2 Natural Processes

Change agents involving natural ecological processes, some of which have been altered by human activities, lead to management challenges associated with maintaining natural processes while managing detrimental effects on sage-grouse habitat conditions, such as invasive species, conifer encroachment, fire, herbivory, weather, and climate. Although there is evidence these have each been altered by human actions, the underlying dynamics of plant growth and propagation, weather, and fire are natural processes and, as such, may require different treatment in research and management compared to specific anthropogenic activities.

4.3.2.1 Conifer Encroachment [H]

The pattern and processes of conifer encroachment have been investigated, and active management of encroachment is underway in many regions. Cost-benefit assessments of conifer treatments in different regions, successional stages, and environmental conditions may help determine the long-term efficacy of these management actions. An understanding of the effects and effectiveness of these management actions in restoring functioning sage-grouse habitat are important factors in this determination. It is important that research addresses appropriate size and location of treatments, characteristics of the target community, effective removal methods that reduce the immediate negative effects of treatment, and the time frame for reuse by sage-grouse populations following treatment. This importance stems in part from the value of woodland habitats to other wildlife, including game species and songbirds (Noson and others, 2006; Anderson and others, 2012).

Interest in management of conifer encroachment stems from two main factors. Conifer trees can provide perching habitat for predators, which could result in increased sage-grouse predation, and conifers displace sagebrush and native herbaceous species. In the modern era, conifer encroachment may result from altered fire regimes (Miller and Rose, 1995, 1999); however biogeographic evidence also indicates long-term, climate-driven trends in juniper woodland expansion (Lyford and others, 2003). Important considerations are the overall consequences of conifer encroachment to sage-grouse habitat relationships and demographic viability, and how long-term changes in conifer abundance have shaped population trends. Management currently focuses on restoring sagebrush habitats by removing trees with as little negative effect on surrounding vegetation and habitat as possible, although, the effects and effectiveness of these treatments on sage-grouse are not understood. For example, the NRCS Sage-Grouse

Initiative is working with private landowners in Oregon, Nevada, and Washington to improve habitat conditions on private lands in regions where conservation success likely depends on private-owner actions. Since its initiation in 2010, projects funded by the Sage-Grouse Initiative have removed more than 115,000 acres of pinyon pine (*Pinus* spp.) and juniper trees from sagebrush communities (Manier and others, 2013, table 27).

More immediately, the effectiveness of tree removal for meeting vegetation, habitat, and wildlife targets is not clear. Some considerations are mechanistic and focus on the condition and successful management of vegetation, for example, determining the most effective control measures for conifer species, and the most effective techniques for restoration of sagebrush and a perennial herbaceous understory in areas with a conifer overstory and depleted sagebrush understory. Other fundamental questions remain about pre- and post-treatment use of these areas by sage-grouse, as well as restoration costs versus benefits. Answers to these questions could help refine methods and improve practices. Addressing these questions requires a full inventory of the distribution and condition of conifer-sagebrush woodlands, including differentiation of stand-ages, description of cover and density of trees and shrubs, and documentation of use and behaviors of sage-grouse in these same regions. Once these are addressed, additional questions can be asked about where conifer encroachment has a negative effect on sage-grouse habitat values, habitat use, and distributions.

4.3.2.2 Invasive Plants [H]

The invasion of the sagebrush landscape by non-native plants has consequences for sage-grouse habitat and may increase fire risk, particularly in cases of widespread cheatgrass (*Bromus tectorum*) infestations. Therefore, the development of new or improved management practices to reduce or eliminate the spread of invasive species is important. Of particular importance is development of methods that eliminate or reduce the distribution and abundance of invasive plants and also promote the re-establishment and productivity of native, herbaceous species. Work by the USDA using natural soil inhibitors (for example, Kennedy and others, 2001) is promising, but test applications are not ready for management implementation. In addition, techniques are needed to restore invaded landscapes to functioning sage-grouse habitat and minimize the risk of reinfestation after treatment. Improved understanding of interactions between invasion history, surface disturbance, habitat condition, and fire history would support the development of these management methods.

Currently (2013), concerns and interests of the wildlife community about invasive plants are focused primarily on annual grasses, including cheatgrass; medusahead wildrye, *Taeniatherum caput-medusae*; and field brome, *Bromus arvense*. Considerations about cheatgrass outweigh all other species. Emphasis on treatments also predominates,

namely discovery and development of treatments to remove cheatgrass without increasing or facilitating its spread. There is recognition that other species can negatively affect habitat conditions, but these threats are not perceived as a current priority. A prudent approach would be to recognize these threats, and develop risk assessments and control options in advance of severe infestations.

In general, habitat managers are seeking integrated invasive-species control methods (for example grazing, mowing, seeding, and herbicides) that minimize negative effects on greater sage-grouse populations and their habitats. Generally, managers recognize that the best techniques may vary by region and local circumstances. Development of improved management practices to minimize the risk of cheatgrass infestation is desired, and progress toward that outcome would benefit from integration and cooperation among managers and researchers. Importantly, the relationship between annual plants and disturbance cycles has been documented (Klemmedson and Smith, 1964; Banks and Baker, 2011; Balch and others, 2012), but practical understanding and control applications are lacking. Research and development needs to focus on interactions between land use, treatment, and disturbance, and emphasize response of vegetation and perpetuation of desirable, perennial species. Adjustment of grazing practices, chemical treatments, biological treatments, physical removal, regional strategies, and regional modeling have been suggested as approaches and applications that warrant consideration in this integrated, adaptive research context.

Although specific applications and treatments are being developed and discovered, developing the information and strategic plans for managing vast acres of public land with perpetual disturbances (creating niches for colonization) and natural variability (affecting restoration and invasion potentials) remain an important consideration. Strategic prioritization, risk assessments, and control approaches that address spatial distributions, seed banks, potential for re-colonization, native species recruitment, and community health have possible value for improving management effectiveness.

4.3.2.2.1 Habitat Condition and Ecosystem Function [H]

The fundamental concerns with invasive plants are the potential and realized detrimental effects on habitat conditions. Besides extreme changes in disturbance regime, as demonstrated by cheatgrass in the Great Basin and Snake River Plain, changes in plant composition can directly affect the forage and cover that sage-grouse require, and also influence the abundance of insects eaten in the spring and during brood rearing (Connelly and others, 2004, 2011a). Therefore, identification of ecological processes and services affected by invasive weeds, with the ability to subsequently prioritize control based on potential for range degradation, would be useful information for habitat management. Potential influences of invasive plants on sage-grouse habitat include loss of native species, altered productivity or palatability, low forage value of invaders, disrupted nutrient cycling, chemical alterations, topsoil erosion, and altered fuel and physical habitat conditions. Functional relationships between invasive plants, disturbances, ecosystem function, ecosystem services, habitat degradation, community resilience, and interactions among ecosystem drivers also need clarification. Interactions between cheatgrass, disturbance, range condition, and probability of wildfire beg better characterization and clarification (Balch and others, 2012). A better explanation of these mechanistic relationships is desired along with more information regarding interactions of climate change with invasive plants and fire regimes.

4.3.2.2.2 Restoration [H]

Research and development of methods for control of invasive plants and restoration of native perennials often is a top priority. This general need (also expressed in section 4.2.1, "Habitat Management, Maintenance, Rehabilitation, and Restoration") encapsulates a suite of specific questions focused on soils, vegetation, and wildlife, as well as disturbance and management treatments to recover or maintain native vegetation. A fundamental issue for restoration includes the determination of optimal seed mixtures appropriate for the soils, climate, and landform of an area. Managers seek information about species to use and seeding practices to prevent reoccurrence of undesirable species. In addition, it is important to consider the effects of variability in species viability and differences between seed sources among regions on restoration success. Long-term control of invasive species, through management of ecological processes, is a key focus for research and development. Important research focuses on the development of methods to prevent non-native plant invasion following fire, reduce threats of short fire-return intervals and related persistence of non-native grassland. A goal of many treatment and restoration projects is to minimize negative short-term effects on habitat quality and distribution and maximize long-term benefits, such as fire prevention, and recovery of shrub and herbaceous productivity. Considerable research and development is needed to develop and evaluate methods to achieve such goals.

4.3.2.3 Fire and Fuels [H]

The threat of large fires and the use of fire-management techniques differ across the range of sage-grouse. Large fires are a primary concern because of the threat they pose to near-term preservation of intact sagebrush ecosystems, and particularly sagebrush cover. A clear understanding of range-wide fire regimes and recovery rates will inform fire management in the sagebrush ecosystem. This information could be obtained, preferably, using historical fire data and by conducting comparative analyses of past fires and fire-related treatments under different environmental conditions. This approach would reduce the need to create new experimental disturbances in intact sagebrush landscapes. Assessments of

fire history and recovery rates could inform planning efforts and deployment of resources for future fire events, improve the understanding of effective post-fire restoration methods, improve the understanding of effectiveness and impacts of pro-active fuel management techniques, and when linked with sage-grouse population and behavioral data, increase the understanding of the response of sage-grouse to fire and fire management.

There are differences in perspectives among regions and management agencies about fire. Clearly, fire is a concern for management and conservation of sagebrush because large fires are a threat to sagebrush cover whenever and wherever fires occur. Further, given current land-cover and land-use patterns, fire is a direct threat to sage-grouse habitats. However, ideas vary about control methods and the uses and ecological roles of fire in sagebrush ecosystems. Fire regimes have a high level of natural variability across the range-wide distribution of sage-grouse (Baker, 2011; Miller and others, 2011). Clear questions relating to fire and habitat quality focus on potential adverse effects, such as reducing large intact habitats, promoting the expansion of invasive species' distributions, and threats to conservation of priority habitats. Nonetheless, fire is understood to be a critical driver of the spatial and temporal dynamics of many Western North American ecosystems, and the understanding of the role of fire in sagebrush ecosystems is incomplete.

There is a clear objection among wildlife research and management communities about using fire to remove more sagebrush, even for research, because sage-grouse and other sagebrush obligates are limited by the distribution of this shrub. In addition, fire creates opportunities for non-native plants. An optional perspective could include fire-related research at local and landscape scales. Continued research addressing the vegetation and faunal response to historical wildfires and previous prescribed-burn treatments is an important link to the adaptive management cycle. Comparative studies assessing and differentiating responses to historical fires and treatments in relation to environmental gradients would provide essential information regarding community development patterns, restoration potential, invasive plant distributions, and wildlife use. Studies and programs that take advantage of recent disturbances through rapid-response plans, and thus, do not require the creation of new disturbances for research, are important.

4.3.2.3.1 Landscape Dynamics and Connectivity [H]

A better understanding of the relationship between fire and the sagebrush ecosystem is important for habitat management and conservation. Attributes associated with this understanding include fire-return intervals, post-fire recovery, fire behaviors, fuel accumulation, ignition sources, frequency of ignition events, and patterns of variation in all of these factors among regions. An improved understanding of these topics requires research addressing historical and current conditions and processes. As historical treatments and wildfires, modern wildfires, and multiple land uses complicate the composition and distribution of sagebrush across the landscape, land-use planners and restoration specialists want a better sense of appropriate or desirable ratios of early, mid- and late-seral communities to provide a guide for balancing habitat values and maintaining productive ecosystems and landscapes. Further, spatial variability in fire history and regimes, suppression and mitigation actions, ecosystem conditions and recovery rates, and associated landscape dynamics add complexity such that multiple, appropriately scaled efforts may be more tractable than range-wide studies. An understanding of the natural and desirable structure and patterns inherent in sagebrush landscapes, and the selection and use patterns of sage-grouse within that structure can directly inform connectivity assessments and habitat conservation.

4.3.2.3.2 Habitat Condition/Effects and Recovery [H]

Identification of the role of fire in protecting and maintaining healthy sagebrush communities is lacking, and identification of the balance between loss of intact sagebrush and recovery of young sagebrush required for habitat maintenance needs to be described for different regions, communities, and ecological types. This would include research directed at informing state-transition pathways and specific practices for improving range conditions and thresholds of degradation. This research would be most acceptable if it did not involve new disturbance of intact sagebrush communities and landscapes, but rather proceeded by taking advantage of existing patterns and previous events whenever possible.

Additional studies addressing the relationship between vegetation conditions before fire and habitat conditions after a fire occurs, including historical, desired, and likely responses, are necessary to understand fire-ecosystem relationships and habitat recovery time. Because of spatial variability, post-fire recovery of sagebrush will be influenced by environmental and biological limitations, historical regimes, and similar determinants of conditions. This situation suggests that implementation of this research regionally, including the range of habitat diversity within and across environmental gradients (for example, McIver and others, 2010), could provide an improved understanding of rangewide post-fire conditions. Further investigation of effects of treatments before fires occur (such as green-stripping) on burn-potentials, habitat conditions, and population demographics would help direct land treatments and enhance the effectiveness of these treatments. A better understanding of the effects of fuel treatments, installation of fire-breaks, and similar manipulation of fuels and (or) landscape patterns to reduce wildfire spread and burn-intensities could help minimize the potential detrimental effects of fire-risk reduction measures.

4.3.2.3.3 Planning and Control Methods [H]

Under extreme fire conditions, managers often have to fight multiple active fires with limited resources. Development of decision-support tools and the research to inform models and prioritization would improve fire-control efforts in sagebrush and sage-grouse habitat. New fire-fighting plans incorporating such information would help with efforts to protect important seasonal sage-grouse habitats, enable prepositioning of resources, and identify strategic locations for fire lines to aid in fire suppression.

4.3.2.3.4 Restoration and Rehabilitation [H]

Large-scale wildfires are likely to continue throughout the Western United States (Ford and others, 2012), and a better understanding of effective habitat rehabilitation following these events can help inform efforts to reduce threats of long-term conversion to non-native grasslands and improve success of rehabilitation efforts. Similarly, development of fuels-management approaches and designs that reduce the threat of large fires in priority habitats and maintain, improve, or minimally affect habitat distribution and quality are needed.

An understanding of spatial variability in ecosystem conditions, fuel profiles, fire history, and response of systems after fire, including sage-grouse populations, can be used to prioritize treatment areas and increase long-term success of management actions. Additionally, protocols for burned-area stabilization and rehabilitation have not been developed for priority sage-grouse habitats. Profiling fuel and erosion risks are just two relevant approaches to defining priorities for restoration projects.

4.3.2.3.5 Vulnerability and Prioritization [H]

Questions remain as to how wildfire should best be managed to minimize detrimental effects on sage-grouse habitat and how to establish priorities for protecting sage-grouse habitat if infrastructure also is in jeopardy. Due to near-term needs for functional (providing food and cover) sagebrush-dominated habitats, focused research on "jump-starting" regeneration of sagebrush and native grasses and forbs to encourage rapid re-establishment of sagebrush and enable re-colonization by sage-grouse as soon as possible may help off-set effects of fire and help mitigate potential effects of future fires. After development, fuel treatment and restoration methods would require testing for effectiveness.

4.3.2.3.6 Population Response [M]

A fundamental component of most sage-grouse research associated with fire is increasing the understanding of how sage-grouse populations respond when seasonal habitats burn. Understanding this response is particularly important when large portions of seasonal habitat burn within a population. Although notable contributions exist from previous work in south-central Idaho (for example, Fischer and others, 1997; Connelly and others, 2000b; Nelle and others, 2000; Pederson and others, 2003), many regional and range-wide perspectives are lacking (but also consider, Knick and others, 2005; Baker, 2011). Changes in sage-grouse demographic rates, including survival and subsequent reproductive success, and movement patterns in response to burned areas, as well as the interactive effects of fire or treatment size and the timing on response of sage-grouse individuals and populations warrant further assessment.

4.3.2.3.7 Interactions with Climate, Grazing, and Other Land Uses [M]

Development of management practices to achieve optimum overstory and understory conditions is an issue because of the complex effects of fire, disturbance, and other land use on these habitat components. Clarifying interactive and determinant effects of climate, fire, and grazing on community composition and ecosystem functions, particularly understory conditions, will inform management for near and long-term ecosystem sustainability, fuels, and habitat management.

4.3.2.4 Herbivory Effects [H]

4.3.2.4.1 Domestic Grazing [H]

Livestock grazing is the most widespread, long-term anthropogenically driven influence on sagebrush ecosystem conditions. It is a contentious issue, and opinions and management approaches vary among regions, States, and localities. Sage-grouse conservation, management of sagebrush ecosystems, and the long-term sustainability of domestic grazing on these lands could benefit from research that informs the relations between grazing practices (for example, intensity, rotation, duration, and other aspects of grazing systems) and regional and local environmental patterns, including soils, climate, fire, and other land uses. Research has provided some insights into relationships between grazing and sage-grouse habitat conditions (Beck and Mitchell, 2000; Beck and others, 2012); however, variability in environmental patterns (for example, climate and vegetation) and grazing practices causes tremendous variability in local and regional effects. An expanded research program would inform immediate decisions and practices to balance grazing practices with conservation practices benefiting wildlife.

Although domestic grazing practices often receive the most attention, wild and domestic herbivores, different types of animals, current and historical conditions, climate, fire, and invasive plants all received notation for their potential interactive effects on habitat conditions. Critical information describing the roles of grazing management in determining ecosystem conditions and the effects on sage-grouse are currently lacking. For example, utilization of forage by elk or wild horses and burros in a drought year may affect availability and use by domestic animals and these interacting uses may further affect sage-grouse habitat conditions. Additionally, research would help refine the timing for re-introduction of livestock after a fire to reduce cheatgrass

invasion risk and provide recovery time for perennial grasses and sagebrush. Considerations could include post-fire condition and relationships between weather and climate and vegetation response.

4.3.2.4.1.1 Practices, BMPs, Systems [H]

There is interest in an improved understanding of how grazing systems, including season of use, grazing duration, kind of livestock, and stocking intensity, influence sage-grouse habitats and populations. A series of large-scale, replicated grazing studies that focus on how different livestock species, grazing systems, disturbance history and other environmental conditions affect sage-grouse habitat would help address these issues and clarify the multitude of conflicting results in the literature.

Any lasting effects of historical overgrazing practices may be distinct from modern practices, and range ecologists may want to separately consider these practices and their effects on sage-grouse habitats. Doing so could improve the ability of managers to address consideration about conditions, for example how "passive" restoration might address effects of current practices Conversely, if historical grazing impacts remain influential, then active restoration may be necessary (Pyke, 2011; Manier and others, 2013). The techniques and associated costs are different for passive and active restoration, and matching methods to desired outcomes is important for cost-effective application.

Specific questions of how fire and habitat treatments interact with grazing interrelate with questions about the effectiveness of post-fire and post-treatment grazing restrictions, and how these treatments and restrictions affect vegetation response and habitat quality. Specific trade-offs between different management approaches have been suggested, such as comparison of the short-term (1–3 years) versus long-term (3–10 years) effects of livestock removal in comparison to best-grazing practices on habitat quality and fire risk (fuels).

4.3.2.4.1.2 Monitoring Effects and Conditions [M]

Development of standardized monitoring protocols to detect trends in vegetation response (vigor, production, and diversity) and similarity to condition, as outlined in the sage-grouse habitat guidelines for addressing the effects of grazing management systems, would be helpful. At a local scale, long-term research and monitoring may have a specific focus, for example to address the response of forbs or the compositional diversity of native species to grazing. At a larger scale, questions may be less specific, such as how grazing regimes affect seasonal sage-grouse distributions from year to year. Both perspectives are necessary to address grazing effects on sage-grouse and the practicality and effectiveness of habitat guidelines in the context of current grazing practices.

Fencing to contain livestock and developments to provide water are common infrastructures associated with grazing. In addition, sometimes sagebrush is removed using mechanical or chemical treatments to improve grazing habitat. Fences may be particularly problematic for sage-grouse, because birds occasionally fly into them and die. Conditions affecting likelihood of collisions and modifications that encourage avoidance have been initiated (for example, Connelly and others, 2004; Wolfe and others, 2009; Stevens and others, 2011, 2012) but the results are not yet adequate to inform and prioritize activities. In general, an understanding of the long-term impacts these infrastructures and treatments to support livestock grazing have on sage-grouse populations would help inform future modification or removal of these features to benefit sage-grouse. Cost-benefit analyses of proposed modifications or removal of infrastructures also would be beneficial.

4.3.2.4.2 Horses and Burros [M]

Some effects of horses and burros on sagebrush ecosystem structure and function have been demonstrated (Beever and Aldridge, 2011). A common assumption is that horses and burros are negatively affecting sage-grouse habitat in the western part of the species' range, but data supporting this assumption are largely lacking. In general, much remains to be learned about the effects of free-ranging horses and burros on sagebrush systems, and how effects vary with equid density and seasonal grazing patterns across ecological contexts and key environmental gradients (for example, rainfall, elevation, seasonality, temperature).

4.3.2.4.3 Wild Herbivores and Herbivore Interactions [L]

Domestic and wild herbivores can occupy the same habitats and their effects can be cumulative across species. For example, interactive effects of multiple herbivores have the potential to exceed independent effects on habitat quality, such as grass and forb abundance and diversity and vegetation structure (Manier and Hobbs, 2007). Furthermore, effects of wild ungulates may disproportionately influence one or more sage-grouse seasonal habitats. Therefore, evaluation of the combined effects of all herbivores is important. A systematic inquiry would include a spatial comparison of key sage-grouse habitats and seasonal habitats for all herbivores and an assessment of impacts on vegetation composition and habitat structure on sage-grouse habitats by herbivores. These analyses would determine degree and timing of spatial overlap and conditions in those overlap zones.

4.3.2.5 Disease [M]

4.3.2.5.1 West Nile Virus [M]

West Nile virus is a local threat to sage-grouse populations when outbreaks occur (Walker and others, 2004, 2007; Walker and Naugle, 2011). Land-use activities, such as oil and gas development, can increase the potential for outbreaks (Walker and others, 2007; Walker and Naugle, 2011). Additional research could examine the effects of other management activities, both as potential sources of West Nile virus and as preventative measures against outbreaks.

A risk assessment for West Nile virus would predict the potential for spread of the virus in different habitat conditions, configurations, and disturbance regimes. Additional research could support the development and testing of methods for vaccination of sage-grouse to protect against West Nile virus infections. Weather conditions appear to have an influence on outbreaks, and the apparent absence of these conditions in recent years complicates efforts to study this disease. Studies would work best if resources were in place to quickly initiate data collection whenever outbreaks occur.

4.3.2.5.2 Background Level of Disease and Implications for Population Cycling [L]

Sage-grouse are susceptible to various diseases (Christiansen and Tate, 2011). Although the impact of disease on populations is presumed to be small, few studies and apparently no ongoing research actually document range-wide background levels of bacterial, fungal, viral, and parasitic diseases. Without these long-term studies, it is difficult to determine if identified disease agents, such as tularemia, aspergillosis, hematozoa, West Nile virus, avian pox, avian malaria, cestodes, coccidian, and other viral and bacterial pathogens, play a role in population cycling or to make prediction of conditions under which outbreaks occur. A range-wide surveillance program would serve as a framework for assessing background disease levels in sage-grouse populations and provide an early warning system for disease outbreaks.

4.3.2.6 Weather and Climate [M]

Under future climate scenarios, the distribution of sage-grouse habitat is predicted to shift as climate and vegetation change (Neilson and others, 2005; Bradley, 2010). Information about the relationship between sage-grouse populations and weather conditions may be particularly important for management as climate patterns shift, causing changes in seasonal patterns and weather variability. In Nevada, certain segments of sage-grouse populations reproduce successfully even under extreme drought conditions (Blomberg and others, 2012a), but it is not clear that all populations can perform similarly. For example, in the Nevada populations, up to 75 percent of annual variability in population size was explained by precipitation (Blomberg and others, 2012a). An understanding of what habitat or landscape features allow populations to survive and reproduce despite climate stresses will help adapt conservation and management plans for future conditions. In general, further study would help characterize the relationship between weather and climate conditions, and timing of use and location of seasonal habitats.

4.3.2.6.1 Implications for Priority Areas [H]

Climate scenarios suggest long-term shifts in the distribution of sage-grouse habitats (Neilson and others, 2005). If this occurs, areas currently designated as sage-grouse priority areas may not serve the same function in the future if sage-grouse habitat no longer occurs there. Assessments are needed to address the adequacy of the current distribution of priority areas to maintain sage-grouse populations under different scenarios of climate change. Additionally, assessment would determine if there are important areas that should receive additional protection as connectivity corridors if or when habitat distributions change. Climate-related research also could help determine if future restoration, mitigation, and rehabilitation efforts should focus on areas predicted to be conducive to supporting future sage-grouse habitat. A high degree of uncertainty about future climate scenarios and high variability in information to address climate-change effects confound research addressing climate change and priority areas.

4.3.2.6.2 Demographics [M]

Seasonal timing of precipitation and temperature may affect vegetation phenology (White and others, 1997; Shen and others, 2011; Friggens and others, 2012) and insect activity (St. Pierre and Lehmkuhl, 1990; Gordo and others, 2010). In Idaho and Utah, sage-grouse chick survival was related to seasonal precipitation and temperature (Guttery and others, 2013). These results highlight the importance of understanding phenological patterns and the development of linkages between these patterns and sage-grouse nest success, adult and juvenile survival, and other processes that affect populations. Further, assessments need to address how birth rate, survival, and mortality and other demographic characteristics are likely to be affected by future climate scenarios and how these patterns influence long-term population viability.

4.3.2.6.3 Cycles and Trends [L]

Sage-grouse populations have undergone population cycling over the past 50 years (Fedy and Doherty, 2011; Garton and others, 2011), and the underlying causes of those cycles are unknown. Climate and weather exhibit multiple cyclic patterns, including drought cycles, oceanic oscillations, and long-term warming (Solomon, 2007). Weather patterns and climate cycles may cause or influence population cycles, although this link has not been established. Further research could assess the cause and effects of these cycles and possible influences on sage-grouse population viability.

4.4 Socio-Economic Considerations

4.4.1 Adaptive Management [M]

There is general agreement that maintaining a tight feedback loop between research and management is important to ensure availability of research results and the use of those results by management. Opportunities for cooperation are abundant, and these efforts may provide feedback regarding implications and effectiveness of local treatments and buffers, as well as measures to achieve regional protections and develop related policies. Understanding and managing the effects of ever-changing forms of industrial developments will require cooperation between planning, management, and research to establish response designs and collect the data necessary to inform decisions.

4.4.2 Economics [M]

Economics and politics are important determinants of policy, potentially affecting land-use developments, habitat protections, and project initiation. Conversely, policies designed to protect sage-grouse potentially restrict economic activities. Therefore, there is value in conducting cost-benefit analyses of the economic impacts of different sage-grouse management options.

4.5 Inventory Data and Products

Data requirements were discussed regularly by participants involved in developing this Research Strategy. Some of the data collections identified emphasize inventory, mapping, or monitoring to acquire data rather than conduct research. Priority data products that could be developed or refined include vegetation composition and community structure, including shrub height, shrub cover, age structure of shrubs, fuel profile, fire frequency, herbaceous cover, productivity of herbaceous vegetation, and exposed mineral soils. Efforts by multiple agencies, such as the BLM Rapid Ecoregional Assessments and FWS Landscape Conservation Cooperatives, are directed toward compilation of data describing infrastructure and surface disturbance, including roads, energy developments, tall structures, and fences. The scale, resolution, and accuracy of these data can be improved as observation technologies advance. Habitat alteration, development, and land use also are ever changing, and ongoing assessment is important to maintain up-to-date data. In addition, continued development, refinement, and downscaling of global and continental scale climate and land-use model projections would improve assessment of landscape change.

Proper data management also is important to maximize use of data that already exist and avoid duplicative data collection. An improved, endorsed, supported, and used means of storing and accessing data related to sage-grouse and sagebrush habitat ecology and management is essential. Modern technologies allow for centralized web storage with a web-interface for data dissemination. Many options exist to develop such a data system, and engagement of the research and management communities is important to ensure that the system works for the sage-grouse user community.

5.0 References Cited

Aldridge, C.L., and Brigham, R.M., 2002, Sage-grouse nesting and brood habitat use in southern Canada: Journal of Wildlife Management, v. 66, p. 433–444.

Aldridge, C.L., Nielsen, S.E., Beyer, H.L., Boyce, M.S., Connelly, J.W., Knick, S.T., and Schroeder, M.A., 2008, Range-wide patterns of greater sage-grouse persistence: Diversity and Distributions, v. 14, p. 983–994.

Anderson, E.D., Long, R.A., Atwood, M.P., Kie, J.G., Thomas, T.R., Zager, P., and Bowyer, R.T, 2012, Winter resource selection by female mule deer *Odocoileus hemionus*— Functional response to spatio-temporal changes in habitat: Wildlife Biology, v. 18, p. 153–163.

Baker, W.L., 2011, Pre-Euro-American and recent fire in sagebrush ecosystems, *in* Knick, S.T., and Connelly, J.W., eds., Greater sage-grouse—Ecology and conservation of a landscape species and its habitats: Berkeley, University of California Press, Studies in Avian Biology, v. 38, p. 185–201.

Balch, J.K., Bradley, B.A., D'Antonio, C.M., and Gomez-Dans, J., 2012, Introduced annual grass increases regional fire activity across the arid western USA (1980–2009): Global Change Biology, v. 19, p. 173–183.

Banks, E.R., and Baker, W.L., 2011, Scale and pattern of cheatgrass (*Bromus tectorum*) invasion in Rocky Mountain National Park: Natural Areas Journal, v. 31, p. 377–390.

Barrows, C.W., Fleming, K.D., and Allen, M.F., 2011, Identifying habitat linkages to maintain connectivity for corridor dwellers in a fragmented landscape: Journal of Wildlife Management, v. 75, no. 3, p. 682–691.

Baumgardt, J.A., Goldberg, C.S., Reese, K.P., Connelly, J.W., Musil, D.D., Garton, E.O., and Waits, L.P., 2013, A method for estimating population sex ratio for sage-grouse using noninvasive genetic samples: Molecular Ecology Resources, v. 13, no. 3, p. 393–402.

Baxter, R. J., Flinders, J.T., Whiting, D.G., and Mitchell, D.L., 2009, Factors affecting nest-site selection and nest success of translocated greater sage grouse: Wildlife Research, v. 36, p. 479–487.

Beck, J.L., Connelly, J.W., and Reese, K.P., 2009, Recovery of greater sage-grouse habitat features in Wyoming big sagebrush following prescribed fire: Restoration Ecology, v. 17, no. 3, p. 393–403.

Beck, J.L., Connelly, J.W., and Wambolt, C.L, 2012, Consequences of treating Wyoming big sagebrush to enhance wildlife habitats: Rangeland Ecology and Management, v. 65, p. 444–455.

Beck, J.L., Reese, K.P., Connelly, J.W., and Lucia, M.B., 2006, Movements and survival of juvenile greater sage-grouse in southeastern Idaho: Wildlife Society Bulletin, v. 34, no. 4, p. 1,070–1,078.

Beck, J.L., and Mitchell, D.L., 2000, Influences of livestock grazing on sage grouse habitat: Wildlife Society Bulletin, v. 28, p. 993–1002.

Beck, T.D.I., and Braun, C.E., 1980, The strutting ground count—Variation, traditionalism, management needs: Proceedings of the Western Association of Fish and Wildlife Agencies, v. 60, p. 558–566.

Beever, E.A., and Aldridge, C.L., 2011, Influences of free-roaming equids on sagebrush ecosystems, with focus on greater sage-grouse, *in* Knick, S.T., and Connelly, J.W., eds., Greater sage-grouse— Ecology and conservation of a landscape species and its habitats: Berkeley, University of California Press, Studies in Avian Biology, v. 38, p. 273–291.

Bell, C.B., and George, T.L., 2012, Survival of translocated Greater Sage-Grouse hens in Northern California: Western North American Naturalist, v. 72, p. 369-376.

Benedict, N.G., Oyler-McCance, S.J., Taylor, S.E., Braun, C.E., and Quinn, T.W., 2003, Evaluation of the eastern (*Centrocercus urophasianus urophasianus*) and western (*Centrocercus urophasianus phaios*) subspecies of sage-grouse using mitochondrial control-region sequence data: Conservation Genetics, v. 4, p. 301–310.

Bennett, A.F., 1999, Linkages in the landscape—The role of corridors and connectivity in wildlife conservation: Cambridge, United Kingdom, IUCN-The World Conservation Union, 254 p.

Bergquist, E., Evangelista, P., Stohlgren, T.J., and Alley, N., 2007, Invasive species and coal bed methane development in the Powder River Basin, Wyoming: Environmental Monitoring and Assessment, v. 128, no. 1–3, p. 381–394.

Berry, J.D., and Eng, R.L., 1985, Interseasonal movements and fidelity to seasonal use areas by female sage grouse: Journal of Wildlife Management, v. 49, p. 237–240.

Blickley, J.L., and Patricelli, G.L., 2012, Potential acoustic masking of greater sage-grouse (*Centrocercus urophasianus*) display components by chronic industrial noise: Ornithological Monographs, v. 74, p. 23–35.

Blickley, J.L., Word, K.R., Krakauer, A.H., Phillips, J.L., Sells, S.N., Taff, C.C., Wingfield, J.C., and Patricelli, G.L., 2012, Experimental chronic noise is related to elevated fecal corticosteroid metabolites in lekking male greater sage-grouse (*Centrocercus urophasianus*): PLoS ONE, v. 7, no. 11, p. e50,462.

Blomberg, E.J., Nonne, D.V., Atamian, M.T., and Sedinger, J.S., 2013, Seasonal reproductive costs contribute to reduced survival of female greater sage-grouse: Journal of Avian Biology, v. 44, p. 149–158.

Blomberg, E.J., Sedinger, J.S., Atamian, M.T., and Nonne, D.V., 2012a, Characteristics of climate and landscape disturbance influence the dynamics of greater sage-grouse populations: Ecosphere, v. 3, p. 55–65.

Blomberg, E.J., Tefft, B.C., Reed, J.M., and McWilliams, S.R., 2012b, Evaluating spatially explicit viability of a declining ruffed grouse population: The Journal of Wildlife Management, v. 76, p. 503–513.

Blus, L.J., Staley, C.S., Henny, C.J., Pendleton, G.W., Craig, T.H., Craig, E.H., and Halford D.K., 1989, Effects of organophosphorus insecticides on sage grouse in southeastern Idaho: Journal of Wildlife Management, v. 53, p. 1,139–1,146.

Bradley, B.A., 2010, Assessing ecosystem threats from global and regional change—Hierarchical modeling of risk to sagebrush ecosystems from climate change, land use and invasive species in Nevada, USA: Ecography, v. 33, p. 198–208.

Braun, C.E., Oedekoven, O.O., and Aldridge, C.L., 2002, Oil and gas development in western North America—Effects on sagebrush steppe avifauna with particular emphasis on sage grouse, in Rahm, J., eds., Transactions of the Sixty-Seventh North American Wildlife and Natural Resource Conference: Washington, D.C., Wildlife Management Institute, p. 337–349.

Brooks, M.L., and Chambers, J.C, 2010, Resistance to invasion and resilience to fire in desert shrublands of North America: Rangeland Ecology and Management, v. 64, p. 431–438.

Bruce, J.R., Robinson, W.D., Petersen, S.L., and Miller, R.F., 2011, Greater sage-grouse movements and habitat use during winter in central Oregon: Western North American Naturalist, v. 71, no. 3, p. 418–424.

Bush, K.L., Dyte, C.K., Moynahan, B.J., Aldridge, C.L., Sauls, H.S., Battazzo, A.M., Walker, B.L., Doherty, K.E., Tack, J., Carlson, J., Eslinger, D., Nicholson, J., Boyce, M.S., Naugle, D.E., Paszkowski, C.A., and Coltman, D.W., 2011, Population structure and genetic diversity of greater sage-grouse (Centrocercus urophasianus) in fragmented landscapes at the northern edge of their range: Conservation Genetics, v. 12, no. 2, p. 527–542.

Carpenter, J., Aldridge, C., and Boyce, M.S., 2010, Sage-grouse habitat selection during winter in Alberta: Journal of Wildlife Management, v. 74, no. 8, p. 1,806–1,814.

Christiansen, T.J., and Tate, C.M., 2011, Parasites and infectious diseases of greater sage-grouse, in Knick, S.T., and Connelly, J.W., eds., Greater sage-grouse–Ecology and conservation of a landscape species and its habitats: Berkeley, University of California Press, Studies in Avian Biology, v. 38, p. 113–126.

Coates, P.S., and Delehanty, D.J., 2010, Nest predation of greater sage-grouse in relation to microhabitat factors and predators: Journal of Wildlife Management, v. 74, p. 240–248.

Colorado greater sage-grouse Steering Committee, 2008, Colorado greater sage-grouse conservation plan, page 2: Colorado Division of Wildlife, accessed July 29, 2013, at http://wildlife.state.co.us/WildlifeSpecies/SpeciesOfConcern/Birds/Pages/GreaterSageGrouseConsPlan2.aspx.

Connelly, J.W., Hagen, C.A., and Schroeder, M.A., 2011b, Characteristics and dynamics of greater sage-grouse populations, in Knick, S.T., and Connelly, J.W., eds., Greater sage-grouse—Ecology and conservation of a landscape species and its habitats: Berkeley, University of California Press, Studies in Avian Biology, v. 38, p. 53–68.

Connelly, J.W., Knick, S.T., Schroeder, M.A., and Stiver, S.J., 2004, Conservation assessment of greater sage-grouse and sagebrush habitats: Cheyenne, Wyoming, Western Association of Fish and Wildlife Agencies, 600 p.

Connelly, J.W., Reese, K.P., Fischer, R.A., and Wakkinen, W.L., 2000b, Response of a sage-grouse breeding population to fire in southeastern Idaho: Wildlife Society Bulletin, v. 20, p. 90–96.

Connelly, J.W., Reese, K.P., and Schroeder, M.A., 2003, Monitoring of greater sage-grouse habitats and populations: Moscow, Idaho, University of Idaho, Natural Resources Experiment Station Report, no. 979.

Connelly, J.W., Rinkes, E.T., and Braun, C.E., 2011a, Characteristics of greater sage-grouse habitats—A landscape species at micro- and macroscales, in Knick, S.T., and Connelly, J.W., eds., Greater sage-grouse—Ecology and conservation of a landscape species and its habitats: Berkeley, University of California Press, Studies in Avian Biology, v. 38, p. 69–83.

Connelly, J.W., Schroeder, M.A., Sands, A.R., and Braun, C.E., 2000a, Guidelines to manage sage grouse populations and their habitats: Wildlife Society Bulletin 28, p. 967–985.

Copeland, H.E., Doherty, K.E., Naugle, D.E., Pocewicz, A., and Kiesecker, J.M., 2009, Mapping oil and gas development potential in the US intermountain West and estimating impacts to species: Plos One, v. 4, no. 10, p. e7,400.

D'Antonio, C.M., and Vitousek, P.M., 1992, Biological invasions by exotic grasses, the grass/fire cycle, and global change: Annual Review of Ecology and Systematics, v. 23, p. 63–87.

Dahlgren, D.K., Chi, R., and Messmer, T.A., 2006, Greater sage-grouse response to sagebrush management in Utah: Wildlife Society Bulletin, v. 34, p. 975–985.

Dahlgren, D.K., Messmer, T.A., and Koons, D.N., 2010a, Achieving better estimates of greater sage-grouse chick survival in Utah: Journal of Wildlife Management, v. 74, no. 6, p. 1,286–1,294.

Dahlgren, D.K., Messmer, T.A., Thacker, E T., and Guttery, M.R., 2010b, Evaluation of brood detection techniques— Recommendations for estimating greater sage-grouse productivity: Western North American Naturalist, v. 70, p. 233–237.

Doherty, K.E., Naugle, D.E., Copeland, H., Pocewicz, A., and Kiesecke, J., 2011, Energy development and conservation tradeoffs—Systematic planning for greater sage-grouse in their eastern range, in Knick, S.T., and Connelly, J.W., eds., Greater sage-grouse—Ecology and conservation of a landscape species and its habitats: Berkeley, University of California Press, Studies in Avian Biology, v. 38, p. 69–83.

Doherty, K.E., Naugle, D.E., Walker, B.L., and Graham, J.M., 2008, Greater sage-grouse winter habitat selection and energy development: Journal of Wildlife Management, v. 72, p. 187–195.

Dzialak, M.R., Olson, C.V., Harju, S.M., Webb, S.L., and Winstead, J.B., 2012, Temporal and hierarchical spatial components of animal occurrence—Conserving seasonal habitat for greater sage-grouse: Ecosphere, v. 3, p. 1–17.

Fedy, B.C., and Aldridge, C.L., 2011, The importance of within-year repeated counts and the influence of scale on long-term monitoring of sage-grouse: Journal of Wildlife Management, v. 75, no. 5, p. 1,022–1,033.

Fedy, B.C., and Doherty, K.E., 2011, Population cycles are highly correlated over long time series and large spatial scales in two unrelated species, greater sage-grouse and cottontail rabbits: Oecologia, v. 165, p. 915–924.

Fedy, B.C., Aldridge, C.L., Doherty, K.E., O'Donnell, M., Beck, J.L., Bedrosian, B., Holloran, M.J., Johnson, G.D., Kaczor, N.W., Kirol, C.P., Mandich, C.A., Marshall, D., McKee, G., Olson, C., Swanson, C.C., and Walker, B.L., 2012, Interseasonal movements of greater sage-grouse, migratory behavior, and an assessment of the core regions concept in Wyoming: The Journal of Wildlife Management, v. 76, p. 1,062–1,071.

Fischer, R.A., Reese, K.P., and Connelly, J.W., 1997, Effects of prescribed fire on movement of female sage-grouse from breeding to summer ranges: Wilson Bulletin, v. 109, p. 82–91.

Ford, P.L., Chambers, J.K., Coe, S.J., and Pendleton, B.C., 2012, Disturbance and climate change in the interior west, in Finch, D.M., ed., Climate change in grasslands, shrublands, and deserts of the interior American West—A review and needs assessment: Fort Collins, Colo., U.S. Forest Service, Rocky Mountain Research Station, RMRS-GTR-285, p. 80–96.

Friggens, M.M., Warwell, M.V., Chambers, J.C., and Kitchen, S.G., 2012, Modeling and predicting vegetation response of western USA grasslands, shrublands, and deserts to climate change, in Finch, D.M., ed., Climate change in grasslands, shrublands, and deserts of the interior American West—A review and needs assessment: Fort Collins, Colorado, United States Forest Service, Rocky Mountain Research Station, RMRS-GTR-285, p. 1–20.

Frye, G.G., Connelly, J.W., Musil, D.D., and Forbey, J.S., 2013, Phytochemistry predicts habitat selection by an avian herbivore at multiple spatial scales: Ecology, v. 94, p. 308–314.

Fundamental Science Practices Advisory Committee, 2011, U.S. Geological Survey Fundamental Science Practices: U.S. Geological Survey Circular 1367, 8 p.

Garton, E.O., Connelly, J.W., Horne, J.S., Hagen, C.A., Moser, A., and Schroeder, M.A., 2011, Greater Sage-Grouse population dynamics and probability of persistence, in Knick, S.T., and Connelly, J.W., eds., Greater sage-grouse— Ecology and conservation of a landscape species and its habitats: Berkeley, University of California Press, Studies in Avian Biology, v. 38, p. 293–381.

Gilpin, M., and Hanski, I., 1991, Metapopulation dynamics— Empirical and theoretical investigations: London, United Kingdom, Academic Press, 336 p.

Gordo, O., Jose-Sanz, J., and Lobo, J.M., 2010, Determining the environmental factors underlying the spatial variability of insect appearance phenology for the honey bee, Apis mellifera, and the small white, Pieris rapae: Journal of Insect Science, v. 10, no. 34.

Gregg, M.A., Dunbar, M.R., and Crawford, J.A., 2007, Use of implanted radiotransmitters to estimate survival of greater sage-grouse chicks: Journal of Wildlife Management, v. 71, p. 646–651.

Gregory, A.J., Kaler, R.S.A., Prebyl, T.J., Sandercock, B.K., and Wisely, S.M., 2012, Influence of translocation strategy and mating system on the genetic structure of a newly established population of island ptarmigan: Conservation Genetics, v. 13, p. 465–474.

Guttery, M.R., Dahlgren, D.K., Messmer, T.A., Connelly, J.W., Reese, K.P., Terletzky, P.A., Burkepile, N., and Koons, D.N., 2013, Effects of landscape-scale environmental variation on greater sage-grouse chick survival: PLoS ONE, v. 8, p. e65,582.

Hagen, C.A., 2011a, Greater sage-grouse conservation assessment and strategy for Oregon—A plan to maintain and enhance populations and habitat: Salem, Oregon Department of Fish and Wildlife, 207 p.

Hagen, C.A., 2011b, Predation on greater sage-grouse—Facts, process and effects, *in* Knick, S.T., and Connelly, J.W., eds., Greater sage-grouse—Ecology and conservation of a landscape species and its habitats: Berkeley, University of California Press, Studies in Avian Biology, v. 38, p. 95–100.

Hanser, S.E., and Knick, S.T., 2011, Greater sage-grouse as an umbrella species for shrubland passerine birds—A multiscale assessment, *in* Knick, S.T., and Connelly, J.W., eds., Greater sage-grouse—Ecology and conservation of a landscape species and its habitats: Berkeley, University of California Press, Studies in Avian Biology, v. 38, p. 475-488.

Hanser, S.E., Aldridge, C.L., Leu, M., Rowland, M.M., Nielsen, S.E., and Knick S.T., 2011, Greater sage-grouse— General use and roost site occurrence with pellet counts as a measure of relative abundance, *in* Hanser, S.E., Leu, M., Knick, S.T., and Aldridge, C.L., eds., Sagebrush ecosystem conservation and management—Ecoregional Assessment Tools and Models for the Wyoming Basins: Lawrence, Kans., Allen Press, p. 112–140.

Hanski, I., 1994, A practical model of metapopulation dynamics: Journal of Animal Ecology, v. 63, p. 151–162.

Harju, S.M., Dzialak, M.R., Taylor, R.C., Hayden-Wing, L.D., and Winstead, J.B., 2010, Thresholds and time lags in effects of energy development on greater sage-grouse populations: Journal of Wildlife Management, v. 74, no. 3, p. 437–448.

Hess, J.E., and Beck, J.L., 2012a, Burning and mowing Wyoming big sagebrush—Do treated sites meet minimum guidelines for greater sage-grouse breeding habitats?: Wildlife Society Bulletin, v. 36, no. 1, p. 85–93.

Hess, J.E., and Beck, J.L., 2012b, Disturbance factors influencing greater sage-grouse lek abandonment in north-central Wyoming: Journal of Wildlife Management, v. 76, no. 8, p. 1,625–1,634.

Holloran, M.J., 2005, Greater sage-grouse (*Centrocercus urophasianus*) population response to natural gas field development in western Wyoming: Laramie, University of Wyoming, Ph.D. dissertation, 215 p.

Holloran, M.J., Heath, B.J., Lyon, A.G., Slater, S.J., Kuipers, J.L., and Anderson, S.H., 2005, Greater sage-grouse nesting habitat selection and success in Wyoming: Journal of Wildlife Management, v. 69, no. 2, p. 638–649.

Holloran, M.J., Kaiser, R.C., and Hubert, W.A., 2010, Yearling greater sage-grouse response to energy development in Wyoming: Journal of Wildlife Management, v. 74, p. 65–72.

Homer, D.G., Aldridge, C.L., Meyer, D.K., Coan, M.J., and Bowen, Z.H., 2009, Multiscale sagebrush rangeland habitat modeling in Southwest Wyoming: U.S. Geological Survey Open-File Report 2008-1027, 14 p.

Homer, D.G., Aldridge, C.L., Meyer, D.K., and Schell, S.J., 2012, Multi-scale remote sensing sagebrush characterization with regression trees over Wyoming, USA—Laying a foundation for monitoring: International Journal of Applied Earth Observation and Geoinformation, v. 14, no. 1, p. 233–244.

Idaho Sage-Grouse Advisory Committee, 2006, Conservation plan for the greater sage-grouse in Idaho: Boise, Idaho Department of Fish and Game, accessed January 29, 2013, at http://fishandgame.idaho.gov/cms/hunt/grouse/conserve_plan/Sage-grousePlan.pdf.

Johnson, D.H., Holloran, M.J., Connelly, J.W., Hanser, S.E., Amundson, C.L., and Knick, S.T., 2011, Influences of environmental and anthropogenic features on greater sage-grouse populations, 1997–2007, *in* Knick, S.T., and Connelly, J.W., eds., Greater sage-grouse—Ecology and conservation of a landscape species and its habitats: Berkeley, University of California Press, Studies in Avian Biology, v. 38, p. 407–450.

Johnson, K.H., and Braun, Johnson, K.H., and Braun, C.E., 1999, Viability and conservation of an exploited sage grouse population: Conservation Biology, v. 13, no. 1, p. 77–84.

Kennedy, A.C., Johnson, B.N., and Stubbs, T.L., 2001, Host range of a deleterious rhizobacterium for biological control of downy brome: Weed Science, v. 49, p. 792–797.

Kirol, C.P., Beck, J.L., Dinkins, J.B., and Conover, M.R., 2012, Microhabitat selection for nesting and brood-rearing by the Greater Sage-Grouse in xeric big sagebrush: Condor, v. 114, no. 1, p. 75–89.

Klemmedson, J.O., and Smith, J.G., 1964, Cheatgrass (*Bromus tectorum L.*): Botanical Review, v. 30, p. 226–262.

Knick, S.T., and Connelly, J.W., eds. 2011, Greater sage-grouse—Ecology and conservation of a landscape species and its habitats: Berkeley, University of California, Studies in Avian Biology, v. 38.

Knick, S.T., and Hanser, S.E., 2011, Connecting pattern and process in greater sage-grouse populations and sagebrush landscapes, *in* Knick, S.T., and Connelly, J.W., eds., Greater sage-grouse—Ecology and conservation of a landscape species and its habitats: Berkeley, University of California, Studies in Avian Biology, v. 38, p. 383–405.

Knick, S.T., Hanser, S.E., and Preston, K.L., 2013, Modeling ecological minimum requirements for distribution of greater sage-grouse leks—Implications for population connectivity across their western range, U.S.A.: Ecology and Evolution, p. 1–13.

Knick, S.T., Holmes, A.L., and Miller, R.F., 2005, The role of fire in structuring sagebrush habitats and bird communities: Studies in Avian Biology, v. 30, p. 63–75.

Larson, M.A., Thompson, F.R., III, Millspaugh, J.J., Dijak, W.D., and Shifley, S.R., 2004, Linking population viability, habitat suitability, and landscape simulation models for conservation planning: Ecological Modelling, v. 180, p. 103–118.

LeBeau, C.W., 2012, Evaluation of greater sage-grouse reproductive habitat and response to wind energy development in south-central, Wyoming: Laramie, University of Wyoming, MS thesis, 138 p.

Leu, M., Hanser, S.E., and Knick, S.T., 2008, The human footprint in the west—A large-scale analysis of anthropogenic impacts: Ecological Applications, v. 18, p. 1,119–1,139.

Lyford, M.E., Jackson, S.T., Betancourt, J.L., and Gray, S.T., 2003, Influence of landscape structure and climate variability on a late Holocene plant migration: Ecological Monographs, v. 73, no. 4, p. 567–583.

Lyon, A.G., and Anderson, S.H., 2003, Potential gas development impacts on sage grouse nest initiation and movement: Wildlife Society Bulletin, v. 31, p. 486–491.

Manier, D.J., and Hobbs, N.T., 2007, Large herbivores in sagebrush steppe ecosystems—Livestock and wild ungulates influence structure and function: Oecologia, v. 152, p. 739–750.

Manier, D.J., Wood, D.J.A., Bowen, Z.H., Donovan, R.M., Holloran, M.J., Juliusson, L.M., Mayne, K.S., Oyler-McCance, S.J., Quamen, F.R., Saher, D.J., and Titolo, A.J., 2013, Summary of science, activities, programs and policies that influence the rangewide conservation of greater sage-grouse (Centrocercus urophasianus): U.S. Geological Survey Open-File Report 2013-1098, 297 p.

McArthur, E.D., and Ott, J.E., 1996, Potential natural vegetation in the 17 conterminous western United States, in Barrow, J.R., McArthur, E.D., Sosebee, R.E., and Tausch, R.J., compilers, Shrubland ecosystem dynamics in a changing environment: U.S. Forest Service General Technical Report INT-GTR-338, p. 16–28.

McCarthy, J.J., and Kobriger, J.D, 2005, Management plan and conservation strategies for greater sage-grouse in North Dakota: Bismark, North Dakota Game and Fish Department, accessed July 30, 2013, at http://gf.nd.gov/gnf/hunting/docs/sage-gr-entire-plan.pdf.

McIver, J.D., Brunson, M., Bunting, S., Chambers, J., Devoe, N., Doescher, P., Grace, J., Johnson, D., Knick, S., Miller, R., Pellant, M., Pierson, F., Pyke, D., Rollins, K., Roundy, B., Schupp, G., Tausch, R., Turner, D., 2010, The Sagebrush Steppe Treatment Evaluation Project (SageSTEP)—A test of state-and-transition theory: U.S. Forest Service General Technical Report, RMRS-GTR-237, 21 p.

Meinke, C.W., Knick, S.T., and Pyke, D.A., 2009, A spatial model to prioritize sagebrush landscapes in the intermountain West (USA) for Restoration: Restoration Ecology, v. 17, no. 5, p. 652–659.

Mezquida, E.T., Slater, S.J., Benkman, C.W., 2006, Sage-Grouse and indirect interactions—Potential implications of coyote control on Sage-Grouse populations: Condor, v. 108, no. 4, p. 747–759.

Miller, R.F., and Rose, J.A., 1995, Historic expansion of Juniperus occidentalis (Western juniper) in southeastern Oregon: Great Basin Naturalist, v. 55, no. 1, p. 37–45.

Miller, R.F., and Rose, J.A., 1999, Fire history and western juniper encroachment in sagebrush steppe: Journal of Rangeland Management, v. 52, no. 6, p. 550–559.

Miller, R.F., Knick, S.T., Pyke, D.A., Meinke, C.W., Hanser, S.E., Wisdom, M.J., and Hild, A.L., 2011, Characteristics of sagebrush habitats and limitations to long-term conservation, in Knick, S.T., and Connelly, J.W., eds., Greater sage-grouse—Eology and conservation of a landscape species and its habitats: Berkeley, University of California Press, Studies in Avian Biology, v. 38, p. 145–184.

Montana Sage Grouse Work Group, 2005, Management plan and conservation strategies for sage grouse in Montana, Final: Helena, Montana Fish, Parks, and Wildlife, accessed July 7, 2013, at http://fwp.mt.gov/fwpDoc.html?id=31187.

Moss, R., Storch, I., and Muller, M., 2010, Trends in grouse research: Wildlife Biology, v. 16, p. 1–11.

Moynahan, B.J., Lindberg, M.S., and Thomas, J.W., 2006, Factors contributing to process variance in annual survival of female greater sage-grouse in Montana: Ecological Applications, v. 16, no. 4, p. 1,529–1,538.

Moynahan, B.J., Lindberg, M.S., Rotella, J.J., and Thomas, J.W., 2007, Factors affecting nest survival of greater sage-grouse in north-central Montana: Journal of Wildlife Management, v. 71, p. 1,773–1,783.

National Technical Team, 2011, A report on national greater sage-grouse conservation measures: Washington, D.C., Bureau of Land Management, 74 p.

Naugle, D.E., Doherty, K.E., Walker, B.L., Copeland, H.E., Holloran, M.J., and Tack, J.D., 2011a, Sage-grouse and cumulative impacts of energy development, *in* Naugle, D.E., ed., Energy development and wildlife conservation in western North America: Washington, D.C., Island Press, p. 55–70.

Naugle, D.E., Doherty, K.E., Walker, B.L., Holloran, M.J., and Copeland, H.E., 2011b, Energy development and greater sage-grouse, *in* Knick, S.T., and Connelly, J.W., eds., Greater sage-grouse—Ecology and conservation of a landscape species and its habitats: Berkeley, University of California Press, Studies in Avian Biology, v. 38, p. 489–503.

Neilson, R.P., Lenihan, J.M., Bachelet, D., and Drapek, R.J., 2005, Climate change implications for sagebrush ecosystems: Transactions of the North American Wildlife and Natural Resources Conference, v. 70, p. 145–159.

Nelle, P.J., Reese, K.P., and Connelly, J.W., 2000, Long-term effects of fire on sage-grouse nesting and brood-rearing habitats in southeast Idaho: Journal of Range Management, v. 53, p. 586–591.

Nevada Sage-Grouse Conservation Team, 2004, Greater sage-grouse conservation plan for Nevada and eastern California: Reno, Nevada Department of Wildlife,accessed July 30, 2013, at http://www.ndow.org/wild/conservation/sg/plan/SGPlan063004.pdf.

Nonne, D.V., Blomberg, E.J., Patricelli, G.L., Krakauere, A.H., Atamian M.T., and Sedinger, J.S., in press, Effects of radio collars on male greater sage-grouse survival and lekking behavior: Condor, v. 115.

Noson, A.C., Schmitz, R.A., and Miller, R.F., 2006, Influence of fire and juniper encroachment on birds in high-elevation sagebrush steppe: Western North American Naturalist, v. 66, p. 343–353.

Noss, R.F., and Peters, R.L., 1995, Endangered ecosystems—A status report on America's vanishing habitat and wildlife: Washington, D.C., Defenders of Wildlife, 132 p.

Oyler-McCance, S.J., and Quinn, T.W., 2011, Molecular insight into the biology of greater sage-grouse, *in* Knick, S.T., and Connelly, J.W., eds., Greater sage-grouse—Ecology and conservation of a landscape species and its habitats: Berkeley, University of California, Studies in Avian Biology, v. 38, p. 85–94.

Oyler-McCance, S.J., St. John, J., 2010, Characterization of small microsatellite loci for use in non invasive sampling studies of Gunnison sage-grouse (*Centrocercus minimus*): Conservation Genetics Resources, v. 2, no. 1, p. 17–20.

Oyler-McCance, S.J., Taylor, S.E., and Quinn, T.W, 2005, A multilocus population genetic survey of the greater sage-grouse across their range: Molecular Ecology, v. 14, no. 5, p. 1,293–1,310.

Patterson, R.L., 1952, The sage grouse in Wyoming: Denver, Colo., Sage Books, 341 p.

Pederson, E.K., Connelly, J.W., Hendrickson, J., and Grant, W.E., 2003, Effect of sheep grazing and fire on sage grouse populations in southeastern Idaho: Ecological Modeling, v. 165, p. 23–47.

Pyke, D.A., 2011, Restoring and rehabilitating sagebrush habitats, *in* Knick, S.T., and Connelly, J.W., eds., Greater sage-grouse—Ecology and conservation of a landscape species and its habitats: Berkeley, University of California Press, Studies in Avian Biology, v. 38, p. 531–548.

Radchuk, V., WallisDeVries, M.F., and Schtickzelle, N., 2012, Spatially and financially explicit population viability analysis of *Maculinea alcon* in the Netherlands: PLoS ONE, v. 7, p. e38,684.

Range-wide Interagency Sage-Grouse Conservation Team, 2012, Near-term greater sage-grouse conservation action plan: Prescott, Ariz., Greater Sage-Grouse Executive Oversight Committee, 27 p.

Reese, K.P., and Connelly, J.W., 1997, Translocations of sage grouse *Centrocercus urophasianus* in North America: Wildlife Biology, v. 3, p. 235–241.

Rowland, M.M., Wisdom, M.J., Suring, L.H., and Meinke, C.W., 2006, Greater sage-grouse as an umbrella species for sagebrush-associated vertebrates: Biological Conservation, v. 129, p. 323–335.

Ryan, M.R., Burger, L.W., Jones, D.P., and Wywialowski, A.P., 1998, Breeding ecology of greater prairie-chickens (*Tympanuchus cupido*) in relation to prairie landscape configuration: The American Midland Naturalist, v. 140, p. 111–121.

Sage-grouse National Technical Team, 2011, A report on national greater sage-grouse conservation measures: U.S. Bureau of Land Management, Washington, D.C., accessed January 29, 2013, at http://www.blm.gov/pgdata/etc/medialib/blm/co/programs/wildlife.Par.73607.File.dat/GrSG%20Tech%20Team%20Report.pdf.

Saunders, D.A., Hobbs, R.J., and Margules, C.R., 1991, Biological consequences of ecosystem fragmentation—A review: Conservation Biology, v. 5, p. 18–32.

Schlaepfer, D.R., Lauenroth, W.K., and Bradford, J.B., 2012, Ecohydrological niche of sagebrush ecosystems: Ecohydrology, v. 5, p. 453–466.

Schroeder, M.A., Aldridge, C.L., Apa, A.D., Bohne, J.R., Braun, C.E., Bunnell, S.D., Connelly, J.W., Deibert, P.A., Garnder, S.C., Hilliand, M.A., Kobriger, G.D., McAdam, S.M., McCarthy, C.W., McCarthy, J.J., Mitchell, D.L., Rickerson, E.V., and Stiver, S.J., 2004, Distribution of sage-grouse in North America: Condor, v. 106, p. 363–376.

Schroeder, M.A., and Baydack, R.K., 2001, Predation and the management of prairie grouse: Wildlife Society Bulletin, v. 29, no. 1, p. 24–32.

Schroeder, M.A., and Vander Haegen, W.M., 2011, Response of greater sage-grouse to the conservation reserve program in Washington State, *in* Knick, S.T., and Connelly, J.W., eds., Greater sage-grouse—Ecology and conservation of a landscape species and its habitats: Berkeley, University of California Press, Studies in Avian Biology, v. 38, p. 517–529.

Schroeder, M., Atamian, A.M., Ferguison, H., Finch, M., Stinson, D.W., 2012, Re-intorduction of sage-grouse to Lincoln County, Washington—2012 Progress Report: Olympia, Washington Department of Fish and Wildlife, 24 p.

Sedinger, B.S., Sedinger, J.S., Espinosa, S., Atamian, M.T., and Blomberg, E.J., 2011, Spatial- temporal variation in survival of harvested greater sage-grouse, *in* Sandercock, B.K., Martin, K., and Segelbacher, G., eds., Ecology, conservation, and management of grouse: Berkeley, University of California, Studies in Avian Biology, v. 39, p. 317–328.

Shen, M., Tang, Y., Chen, J., Zhu, X. and Zheng, Y., 2011, Influences of temperature and precipitation before the growing season on spring phenology in grasslands of the central and eastern Qinghai-Tibetan Plateau: Agricultural and Forest Meteorology, v. 151, no. 12, p. 1,711–1,722.

Solomon, S., Qin, D., Manning, M., Chen, Z., Marquis, M., Averyt, K.B., Tignor, M., and Miller, H.L., eds., 2007, Climate change 2007—The physical science basis, contribution of working group I to the fourth assessment report of the intergovernmental panel on climate change: New York, Cambridge University Press, 996 p.

South Dakota Department of Game, Fish, and Parks, 2008, Greater sage-grouse management plan South Dakota 2008-2017: Pierre, South Dakota Department of Game, Fish, and Parks, accessed January 29, 2013, at http://gfp.sd.gov/wildlife/docs/sage-grouse-management-plan.pdf.

Stevens, B.S., Reese, K.P., and Connelly, J.W., 2011, Survival and detectability bias of avian fence collision surveys in sagebrush steppe: Journal of Wildlife Management, v. 75, no. 2, p. 437–449.

Stevens, B.S., Reese, K.P., Connelly, J.W., and Musilb, D.D., 2012, Greater sage-grouse and fences—Does marking reduce collisions?: Wildlife Society Bulletin, v. 36, p. 297–303.

Stinson, D.W., Hays, D.W., and Schroeder, M.A., 2004, Washington State recovery plan for the greater sage-grouse: Olympia, Washington Department of Fish and Wildlife, 109 p., accessed July 30, 2013, at http://wdfw.wa.gov/publications/00395/wdfw00395.pdf.

Stiver, S.J., Apa, A.D., Bohne, J.R., Bunnell, S.D., Deibert, P.A., Gardner, S.C., Hilliard, M.A., McCarthy, C.W., and Schroeder, M.A., 2006, Greater sage-grouse comprehensive conservation strategy: Cheyenne, Wyo., Western Association of Fish and Wildlife Agencies, accessed January 29, 2013, at http://www.wafwa.org/pdf/GreaterSage-grouseConservationStrategy2006.pdf.

Stiver, S.J., Rinkes, E.T., and Naugle, D.E., 2010, Sage-grouse habitat assessment framework: U.S. Bureau of Land Management, Idaho State Office, Boise, Idaho, 135 p., accessed July 30, 2013, at http://www.blm.gov/pgdata/etc/medialib/blm/wo/Communications_Directorate/public_affairs/sage-grouse_planning/documents.Par.23916.File.dat/SG_HABITATASESSMENT0000669.pdf.

St. Pierre, R.G., and Lehmkuhl D.M., 1990, Phenology of *Hoplocampa montanicola* Rohwer (Tenthredinidae) and *Anthonomus quadrigibbus* Say (Curculionidae) on their host plant *Amelanchier alnifolia* Nutt. (Rosaceae) in Saskatchewan: Canadian Entomologist, v. 122, p. 901–906.

Sveum, C.M., Crawford, J.A., and Edge, W.D., 1998a, Use and selection of brood-rearing habitat by sage grouse in south central Washington: Great Basin Naturalist, v. 58, p. 344–351.

Sveum, C.M., Crawford, J.A., and Edge, W.D., 1998b, Nesting habitat selection by sage grouse in south-central Washington: Journal of Range Management, v. 51, p. 265–269.

Tack, J.D., 2009, Sage-grouse and the human footprint—Implications for conservation of small and declining populations: Missoula, Montana, University of Montana, M.S. thesis, p. 96.

Tack, J.D., Naugle, D.E., Carlson, J.C., and Fargey, P.J., 2012, Greater sage-grouse (*Centrocercus urophasianus*) migration links the USA and Canada—A biological basis for international prairie conservation: Oryx, v. 46, no. 1, p. 64–68.

Taylor, R.L., Walker, B.L., Naugle, D.E., and Mills, L.S., 2012, Managing multiple vital rates to maximize greater sage-grouse population growth: Journal of Wildlife Management, v. 76, no. 2, p. 336–347.

U.S. Department of Energy, 2008, 20% wind energy by 2030—Increasing wind energy's contribution to US electricity supply: Washington, D.C., U.S. Department of Energy, 248 p.

U.S. Department of the Interior, 2005, Endangered and threatened wildlife and plants—12- month finding for petitions to list the greater sage- grouse as threatened or endangered—Proposed rule: Federal Register, v. 70, p. 2,244–2,282.

U.S. Department of the Interior, 2010, Endangered and threatened wildlife and plants—12-month findings for petitions to list the Greater Sage- Grouse (*Centrocercus urophasianus*) as threatened or endangered: Federal Register, v. 75, p. 13,910–13,958.

U.S. Department of the Interior, 2012, Sage-grouse conservation objectives draft report: United States Department of the Interior, accessed July 30, 2013, at http://www.fws.gov/mountain-prairie/species/birds/ grouse/20120803ConservationObjectivesTeamDraftReport. pdf.

U.S. Fish and Wildlife Service, 2013, Greater sage-grouse (*Centrocercus urophasianus*) conservation objectives—Final report: Denver, Colo., U.S. Fish and Wildlife Service, 115 p.

U.S. Geological Survey, 2008, Landcover map: U.S. Geological Survey Northwest Gap Analysis Project, accessed July 30, 2013, at http://gapanalysis.usgs.gov.

U.S. Geological Survey, 2011, LANDFIRE 1.1.0 Existing Vegetation Type layer: U.S. Department of the Interior, Geological Survey, accessed July 30, 2012, at http://landfire. cr.usgs.gov/viewer/.

Utah Division of Wildlife Resources, 2009, Utah greater sage- grouse management plan: Salt Lake City, Utah Department of Natural Resources, Division of Wildlife Resources, Publication 09-17, 94 p.

Utah Wildlife in Need, 2011, Protocol for investigating the effects of tall structures on sage-grouse within designated and proposed energy corridors: Utah Wildlife in Need, 50 p., accessed July 31, 2013, at http://www.utahcbcp.org/files/ uploads/UWIN_SageGrouse_Structure_ProtocolFinal.pdf.

Veith, M., and Schmitt, T., 2008, Conservation genetics: potential and limitations: Zeitschrift fuer Feldherpetologie, v. 15, no. 2, p. 119–138.

Walker, B.L., Naugle, D.E., Doherty, K.E., and Cornish, T.E., 2004, From the field—Outbreak of West Nile virus in greater sage-grouse and guidelines for monitoring, handling, and submitting dead birds: Wildlife Society Bulletin, v. 32, no. 2, p. 1,000-1,006.

Walker, B.L., Naugle, D.E., and Doherty, K.E., 2007, Greater sage-grouse population response to energy development and habitat loss: Journal of Wildlife Management, v. 71, no. 8, p. 2,644–2,654.

Walker, B.L., and Naugle, D.E., 2011, West Nile virus ecology in sagebrush habitat and impacts on greater sage-grouse populations, *in* Knick, S.T., and Connelly, J.W., eds., Greater sage-grouse—Ecology of a landscape species and its habitats: Berkeley, University of California Press, Studies in Avian Biology, v. 38, p. 127–143.

Walsh, D.P., White, G.C., Remington, T.E., and Bowden, D.C., 2004, Evaluation of the lek-count index for greater sage- grouse: Wildlife Society Bulletin, v. 32, p. 56–68.

Watters, M.E., McLash, T.L., Aldridge, C.L., and Brigham, R.M., 2002, The effect of vegetation structure on predation of artificial greater sage-grouse nests: Ecoscience, v. 9, no. 3, p. 314–319.

White, M.A., Thornton, P.E., and Running, S.W., 1997, A continental phenology model for monitoring vegetation responses to interannual climatic variability: Global Biogeochemical Cycles, v. 11, no. 2, p. 217–234.

Wisdom, M.J., Rowland, M.M., Hemstrom, M.A., Wales, B.C., 2005a, Landscape restoration for greater sage-grouse— Implications for multiscale planning and monitoring, *in* Shaw, N.L, Pellant, M., and Monsen, S.B., comps., Sage- Grouse Habitat Restoration Symposium Proceedings: Sage- Grouse Habitat Restoration Symposium, Boise, Idaho, 2001 [Proceedings], p. 62–69.

Wisdom, M.J., Rowland, M.M., and Suring, L.H., eds, 2005b, Habitat threats in the sagebrush ecosystem—Methods of regional assessment and applications in the Great Basin: Lawrence, Kansas, Alliance Communications Group.

Wisdom, M.J., Meinke, C.W., Knick, S.T., and Schroeder, M.A., 2011, Factors associated with extirpation of Sage- Grouse, *in* Knick, S.T., and Connelly, J.W., eds., Greater sage-grouse—Ecology of a landscape species and its habitats: Berkeley, University of California Press, Studies in Avian Biology, v. 38, p. 451–472.

Wolfe, D.H., Patten, M.A., and Sherrod, S.K., 2009, Reducing grouse collision mortality by marking fences: Ecological Restoration, v. 27, p. 141–143.

Wyoming Sage-Grouse Working Group, 2003, Wyoming greater sage-grouse conservation plan, Wyoming Game and Fish Commission: Cheyenne, Wyo., Western Association of Fish and Wildlife Agencies, accessed July 30, 2013, at http://wgfd.wyo.gov/web2011/Departments/Wildlife/pdfs/ SG_WGFDFINALPLAN0000653.pdf.

Appendix A.   Research Questions Identified from a Review of Federal and State Sage-Grouse Conservation Document, Peer-Reviewed Papers, and Input from the Scientific Community

Questions have been categorized into a hierarchical organizational structure by theme and topics addressed. The PDF (portable document format) file can be downloaded at http://pubs.usgs.gov/sir/2013/5167/.